The Ten Commandments Kinematically Aligned Total Knee Arthroplasty:

A Primer for the Orthopedic Surgeon and an Introduction for the Discerning Patient
Color Edition

The modern field of total knee arthroplasty began over 50 years ago in the 1970s when the FDA approved cement for implant fixation, and durable plastic that replaced the meniscus became available. Simultaneously, alignment evolved into two opposing and contentious philosophies, the anatomic approach (forefather of kinematic alignment,) and the 'cookie-cutter' or mechanical alignment.

In 2006, the concept of personalized surgery initiated a paradigm shift away from the philosophy of mechanical alignment to calipered kinematic alignment total knee replacement. Dr. Howell, a professor of biomedical engineering and a sports medicine surgeon, developed the kinematic alignment technique. To perform the surgery, he created the first commercially made patient-specific guides designed to assist the surgeon in setting the components coincident to the patient's pre-arthritic joint lines, which is the kinematic alignment target.

In 2007, early adopters' clinical experience, particularly Dr. Meade, persuaded over 300 surgeons to perform more than 20,000 kinematically aligned total knee replacements by 2011. Between 2011 and 2020, many worldwide studies reported that kinematic alignment improves patient satisfaction, function, ease of recovery, soft tissue balance, flexion, and joint-line and limb alignment compared with mechanical alignment. It is a winning approach.

This primer uses the familiar concept of the Ten Commandments and selects ten topics for the surgeon to follow. The success of kinematic alignment requires caliper measurements of bone resections and intraoperative recording of verification steps. These recommendations reduce the risk of complications, a topic of great interest to the patient!

Patients that research options in total knee replacement surgery will come across the kinematic and mechanical alignment philosophies. The mechanical alignment technique is a 'cookie-cutter' approach that places the components the same way in everybody regardless of their legs' shape. Because the method changes the patient's joint lines, the healthy ligaments are released, making the replaced knee feel unnatural. Kinematic alignment restores the patient's knee alignment before arthritis developed and preserves healthy ligaments, promoting patient satisfaction and high function.

We hope this primer is useful for those investigating knee replacement surgery. Those who chose kinematic alignment, ask your surgeon whether they use the caliper and intraoperative verifications to position your new knee optimally!

Thomas D. Meade M.D.
Stephen M. Howell M.D

The Ten Commandments of
Calipered Kinematically Aligned Total Knee Arthroplasty:

*A Primer for the Orthopedic Surgeon and
an Introduction for the Discerning Patient Ten Commandments*

Color Version

Copyright © 2021

Thomas D. Meade M.D. & Stephen M. Howell, M.D.

ISBN Information: The International Standard Book Number (ISBN) is a unique machine-readable identification number, which marks any book unmistakably. The ISBN is the clear standard in the book industry. One hundred fifty-nine countries and territories are officially ISBN members.

ISBN: 978-1-961879-00-3 (sc)
978-1-961879-01-0 (e)

Publishing rev. date: 08/06/2023

AUTHOR CABIN
THE PLACE FOR YOUR STORY

Dedication

We dedicate this primer to those pioneering surgeons who, 50 years ago, dreamed of changing people's lives by developing one of the most important orthopedic surgical advances of the 20th century- Total Knee Replacement. And to the patient who teaches the orthopedic surgeon, we are indebted to the selfless sharing of their clinical experiences, which is fundamental for scrutinizing today's techniques and improving total knee replacement tomorrow.

Acknowledgments:

Additionally, we thank Cheryl Perugini, who painstakingly listened to every word of Dr. Howell's VuMedi 10 Commandment presentations multiple times to 'translate' casual medical-speak into readable prose.

Table of Contents

Preface

For the Orthopedic Surgeon:

The modern field of total knee arthroplasty began over 50 years ago. In the 1970s, the combined effect of FDA approval of cement for implant fixation (i.e., methylmethacrylate) and the availability of durable plastic as a tibial bearing (i.e., ultra-high molecular weight polyethylene) fostered innovation of knee replacement implant designs. Simultaneously, alignment evolved into two opposing and contentious philosophies, the anatomic approach and the mechanical alignment approach, which is the focus of this primer.

The 'personalized' or 'anatomic' alignment philosophy proposed by David Hungerford, MD from Johns Hopkins University, should be considered the predecessor of calipered kinematically aligned total knee replacement. Hungerford developed universal instruments based on the concept of measured resection so that the bone and cartilage removed equaled the thickness of the components. Equal thickness set the components coincident to the patient's pre-arthritic joint lines and retained healthy ligaments, preserving the kinematics of the native knee. Restoring the patient's pre-arthritic joint lines, which co-aligns the three kinematic axes of the knee with the components' three rotational axes is the foundation of calipered kinematically aligned total knee replacement.

In contrast, the 'cookie-cutter' or mechanical alignment philosophy proposed by John Insall, MD from the Hospital for Special Surgery in New York City, chose to overlook individual differences by cutting the femur and tibia at a right angle to the long axis of the femur and tibia in all knees. One popular version of mechanical alignment involves establishing a rectangular-shaped extension and flexion spaces with medial and lateral gaps of equal laxity throughout the motion arc. Obtaining gaps of similar laxity requires the release of healthy ligaments with imprecise and non-reproducible techniques, which tightens and misshapes the patient's pre-arthritic flexion space. No mechanical alignment version co-aligns the components' three rotational axes with the kinematic axes and they routinely cut healthy ligaments. Morbidity from mechanical alignment's multiple ligament' injuries' and component deviations from the kinematic axes cause pain, stiffness, and instability and slow the patient's recovery.

The mechanical alignment philosophy, combined with newer implant designs and technologies proffering more accurate alignment, provided little improvement in patient satisfaction by the early 2000s. Case-series from multiple skilled arthroplasty surgeons and international joint registries reported a level of dissatisfaction in one out of 5 patients with a mechanically aligned TKA, providing an impetus for a paradigm change in alignment philosophy.

In 2006, the tenet of personalized surgery caused a paradigm shift away from mechanical alignment to calipered kinematically aligned total knee replacement. Dr. Howell, a professor of biomedical engineering and sports medicine surgeon, developed the kinematic alignment technique and the first commercially made patient-specific guides designed to assist the surgeon in setting the components coincident to the patient's pre-arthritic joint lines. In 2007, communication of the clinical experience of early adopters, especially Dr. Meade, persuaded over 300 surgeons to perform more than 20,000 kinematically aligned total knee replacements by 2011. Between 2011 and 2020, many worldwide studies reported kinematic alignment improves patient satisfaction, function, ease of recovery, soft tissue balance, flexion, and joint-line and limb alignment compared with mechanical alignment.

`This primer uses the familiar concept of the Ten Commandments. It selects ten topics designed to educate the surgeon about the technique of kinematic alignment and the critical use of caliper measurements of bone resections. The intraoperative recording of verification steps reduces the risk of complications, a topic of great interest to the patient!

For Our Patients:

When researching total knee replacement surgical options, you will come across the kinematic alignment and mechanical alignment philosophies.

What's the difference?

Mechanical alignment is a 'cookie-cutter' technique aligning the components the same way in all patients. Unfortunately, because our knees and bodies are unique in size, shape, and alignment, a mechanically aligned total knee replacement often changes the natural alignment of both your knee and your leg itself. Because of this, 20-25% of patients with mechanically aligned total knee replacements complain of pain, stiffness, and instability.

Kinematic alignment is a personalized technique that adjusts caliper measurements of bone and cartilage pieces to match the components' thickness and naturally align your leg and knee. Restoring your natural alignment increases the likelihood that the artificial knee will feel and function like your original knee.

Patients treated with a kinematically aligned total knee replacement report better pain relief, function, and bending of the knee than patients who chose a mechanical knee replacement. They notice a less painful and quicker recovery, as many patients walk without a cane and drive a car within four weeks of surgery. The caliper's use to measure and adjust the components' position and orientation to within ± 0.5 millimeters of the patient's pre-arthritic joint lines explains these improvements. The caliper is much more accurate, quicker, and has fewer complications than robotics and navigational instruments.

Kinematically aligned knee replacements are a minimally invasive surgical (MIS) procedure with a low risk of infection and complications. Because your knee ligaments are not released as they often are in mechanical alignment, the risk of a blood transfusion is negligible.

The kinematically aligned total knee replacement surgery itself is short and typically takes less than 50 minutes. And 90% of the time, the patient can be discharged on the day of surgery, depending on their general health.

Finally, the implanted components should function along time (more likely longer than you) without another surgery. Implant survival at three, six, and up to nine years is equivalent or better than mechanical alignment knee replacements.

When it comes to having a total knee replacement, the choice of alignment should be clear; however, there's one more thing you'll need to figure out: finding the right surgeon.

While some orthopedic surgeons have experience with mechanically aligned knee replacements, they may not have learned the calipered kinematic alignment method. Be an educated consumer, do your homework before embarking on the knee replacement journey.

Make sure your surgeon uses the caliper and intraoperative verifications to kinematically align your new knee optimally!

About the Author
Thomas D. Meade, M.D.

Thomas Meade, MD serves as Department Chair, Orthopedic Knee Arthroplasty with Coordinated Health/Lehigh Valley Health Network and has held academic positions at The Commonwealth Medical College, DeSales University and the Hershey Medical College of The Pennsylvania State University.

Dr Meade received his medical degree from Jefferson Medical College with AOA honors after graduating from The Pennsylvania State University with High Distinction. He completed his orthopaedic residency at Thomas Jefferson University/Rothman Institute where he received the outstanding senior research award, followed by a fellowship in sports medicine and advanced knee reconstruction with Dr Frank Noyes in Cincinnati.

Dr Meade has practiced in NE Pennsylvania since 1989 and has performed more than 18,000 procedures, including 15 years of Level 1 trauma surgery. He designed and developed one of the largest orthopaedic facilities in the country, in addition to developing a large sports medicine company caring for dozens of high schools, colleges, Olympic and professional teams. The Philadelphia Eagles and Philadelphia Flyers have retained Dr. Meade's services as a consultant.

He is an experienced clinician in the field of knee replacements, knee ligament reconstruction and has limited his practice to knee surgery since 2004. He holds numerous patents for orthopaedic implants, instruments, and techniques. He has served as consultant for many major orthopaedic implant companies including surgeon consultant-designer for the 2nd generation iBalance Knee-Arthroplasty System. He has authored numerous articles, book chapters, books and is the host of two locally syndicated TV shows: Real Life in The OR and The Dr Meade Show (SSPTV.com)

He speaks and consults internationally and is active in clinical research in knee arthroplasty, computer generated 3-D guides, nutriceuticals in addition to being a national leader in outpatient total knee arthroplasty, kinematic alignment knee arthroplasty and opioid sparing orthopaedic surgery.

Dr. Meade received the 2002 Founders Award from the prestigious Medical Fitness Association for his lifetime contributions to the field of medical fitness, the Sir John Charnley Award for orthopedic achievement, 11 state and national government citations and commendations for orthopaedic excellence and 4 times Best Surgeon-People's Choice award.

Dr Meade is an avid cyclist, a competitive swimmer and triathlete and has qualified for the Ironman Triathlon. He has held top ten individual US Masters Swimming rankings and several USMS world relay records.

About the Author
Stephen M. Howell, M.D.

Dr. Howell is an orthopedic surgeon practicing in Sacramento and Lodi California and is an Adjunct Professor of Biomedical Engineering at the University of California, Davis. His work improving the understanding of total knee replacement, anterior cruciate ligament reconstruction, and surgical treatment of meniscus lesions is based on a 26-year collaboration with Distinguished Professor Maury L. Hull, Ph.D. Drs. Howell and Hull have collectively educated and graduated four students with Ph.D. degrees and 30 students with Master's degrees in mechanical or biomedical engineering. Dr. Howell has published more than 184 scientific articles in peer-reviewed journals.

Dr. Howell frequently shares his clinical experiences and research findings as an invited speaker at national and international meetings and universities worldwide. He regularly hosts experienced orthopedic surgeons worldwide who seek education in his techniques by observing patient care in the office, hospital, and operating room. He is a consultant for major orthopedic implant companies worldwide. Dr. Howell developed instrumentation, implants, and educational materials for Medacta, ThinkSurgical, and Zimmer Biomet to help surgeons perform calipered kinematically aligned total knee replacements. He is also a consultant for Zimmer Biomet Sports Medicine and has developed instrumentation, implants, and educational materials for ACL reconstruction, which national and international surgeons use. Dr. Howell holds 31 U.S. patents and 3 European patents.

A member of many prestigious professional societies, Dr. Howell is an Honorary Member of the German Arthroplasty Society, a past-president of the International ACL Study Group, and a member of the American Association of Hip and Knee Surgeons, the International Society of Arthroscopy, Knee Surgery and Orthopedic Sports Medicine, the American Academy of Orthopedic Surgeons, as well as the Arthroscopy Association of North America. He is an Emeritus Editor of the American Journal of Sports Medicine.

A 1981 graduate of Northwestern University's Feinberg School of Medicine, Dr. Howell completed his internship at the Graduate Hospital of the University of Pennsylvania in Philadelphia and his residency in orthopedic surgery at Philadelphia's Thomas Jefferson University Hospital in 1986. Before entering private practice in 1989, Dr. Howell served in the Air Force as an active duty orthopedic surgeon at Travis Air Force Base, where he earned the rank of Lieutenant Colonel. He completed 14 additional years as a reservist, which included a recall to active duty for Gulf War I.

Chapter 0   Overview of Ten Commandments

This overview provides the biomechanical foundation for following the Ten Commandments by describing the three kinematic axes location and the importance of using a caliper to co-align them with the axes of the femoral and tibial components. Doing so will reduce your risk of complications, improve patient satisfaction, build your practice, and make your life easier.

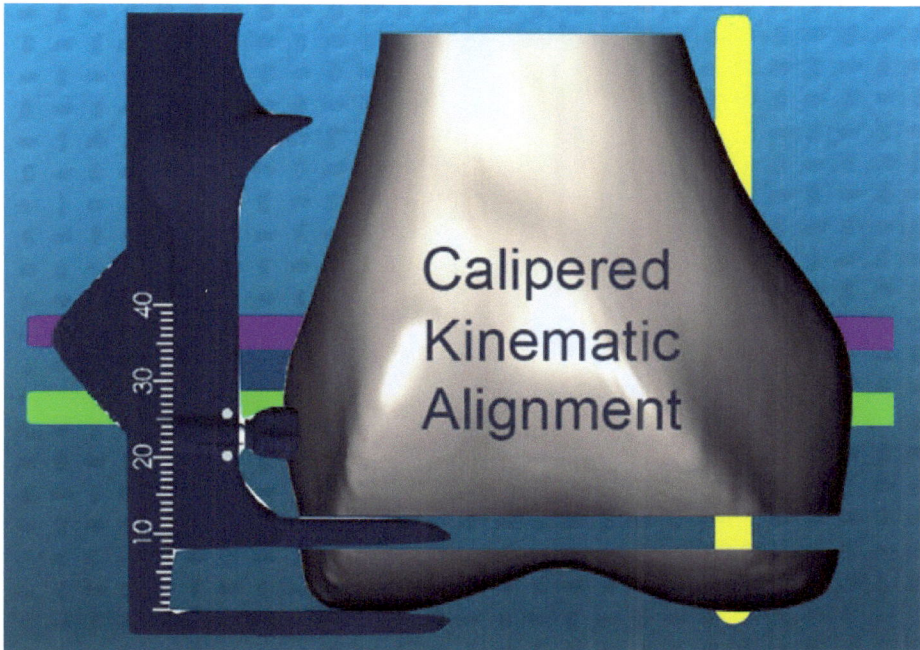

What is the definition of Calipered Kinematic Alignment?

Kinematic alignment restores the patient's pre-arthritic joint lines and three kinematic axes without releasing healthy ligaments. It does so by using a caliper to measure the thickness of bone resections and then adjust the thickness to within ± 0.5 mm of the thickness of the distal and posterior condyles of the femoral component after compensating for cartilage wear and the thickness of the kerf or bone lost from the saw cut.

This schematic, comprised of four poses of the distal femur, shows the orthogonal relationship between these kinematic axes and the distal and posterior femoral joint lines.

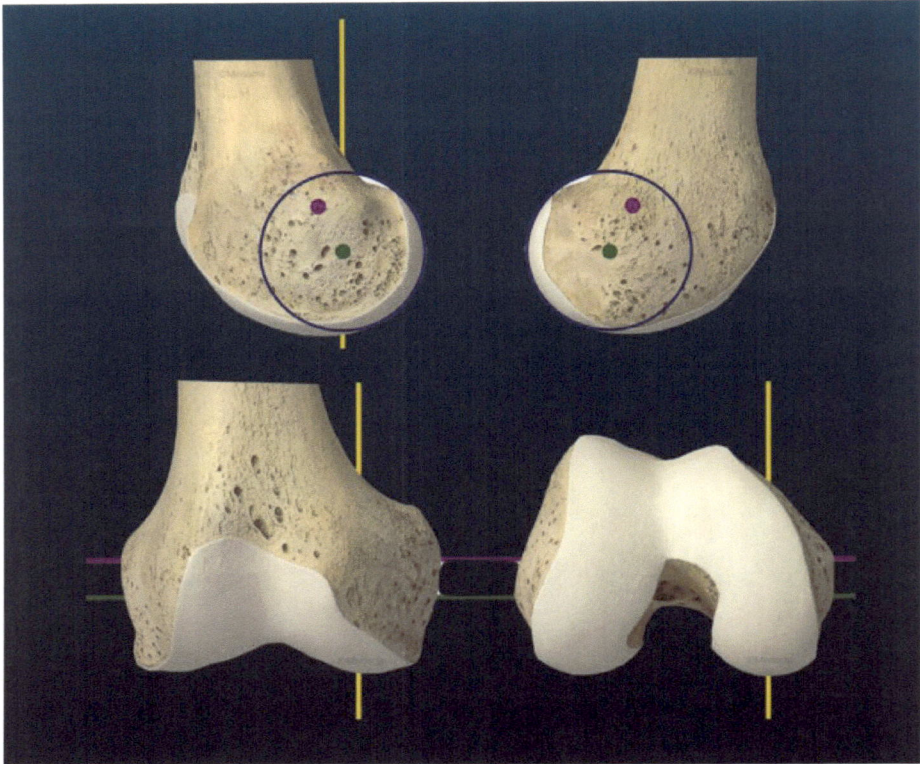

The green line is an axis that connects the centers of best fit circles applied to the medial and lateral femoral condyle.

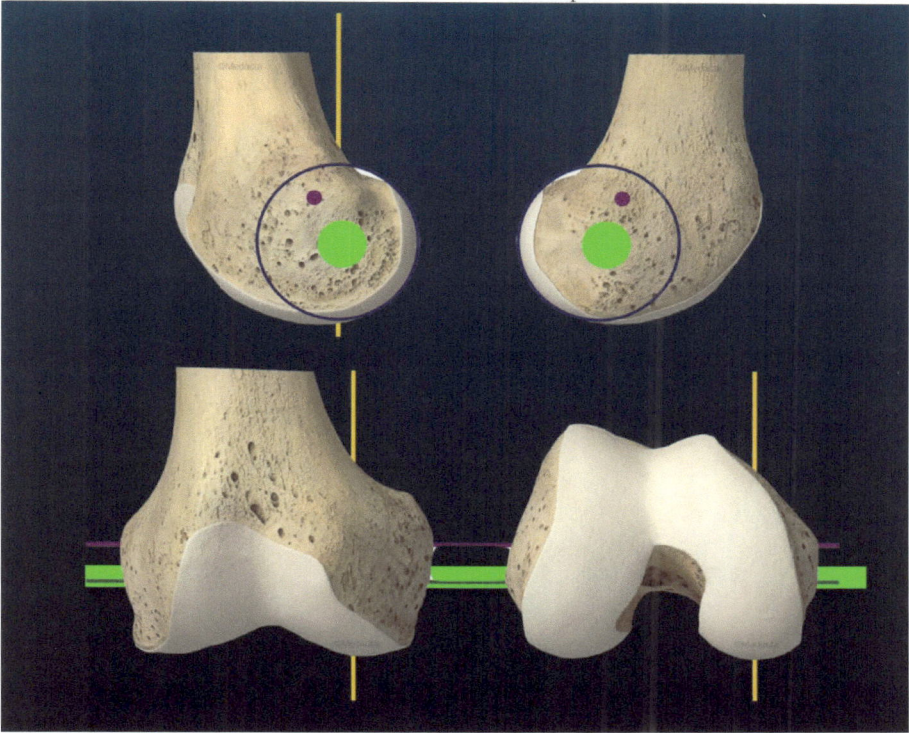

The femur's green axis is the one about which the tibia flexes and extends and is parallel to the patient's pre-arthritic distal and posterior femoral joint line.

The magenta line is an axis that is approximately 10 to 12 millimeters anterior, proximal, and parallel to the first one.

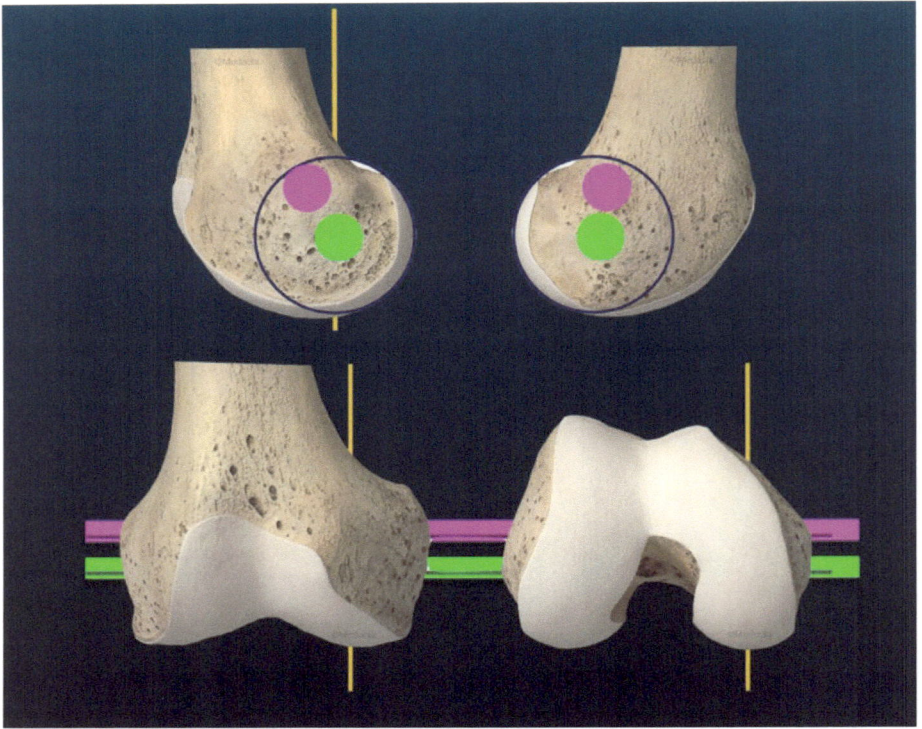

The femur's magenta axis is the one about which the patella flexes and extends and is parallel to the patient's pre-arthritic distal and posterior femoral joint line and the green axis.

The yellow line is a longitudinal or vertical axis slightly posterior to the center of the medial compartment.

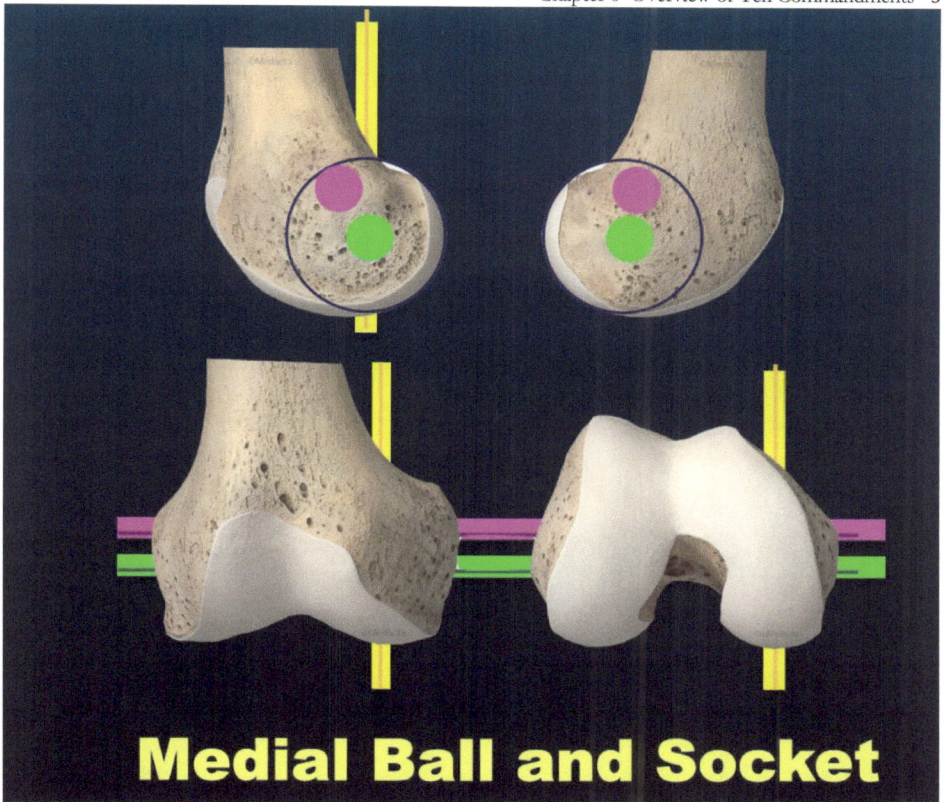

Medial Ball and Socket

The femur's longitudinal yellow axis is perpendicular to the patient's pre-arthritic distal and posterior femoral joint line and the green and magenta axes.

These three axes are the only axes in the knee, and they define native knee kinematics. When the femoral component is not coincident with the patient's pre-arthritic joint lines, the components' axes are not co-aligned with kinematic axes. The deviations between the axes can tighten and slacken the natural resting length of the healthy collateral, retinacular, and posterior cruciate ligaments causing stiffness and instability.

When the components resurface, the axes are co-aligned with the femoral and tibial components, creating 'kinematic harmony' between the components and soft tissues.

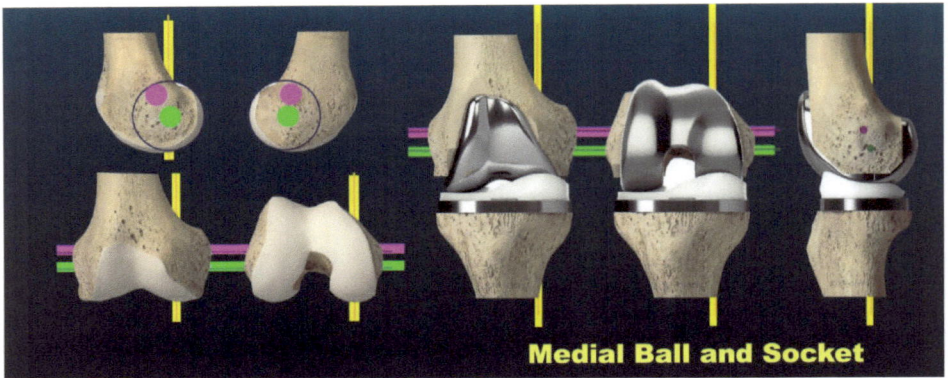

Medial Ball and Socket

This schematic shows that setting the femoral and tibial components coincident to the patient's pre-arthritic joint lines co-aligns the three kinematic axes of the native knee with those of the components. Co-aligning the axes allow the knee to work like a medial ball-in-socket, which replicates native knee kinematics.

Four measurements set the frontal component coincident to the patient's pre-arthritic joint line that requires the use of the essential intraoperative instrument for total knee replacement--- the caliper.

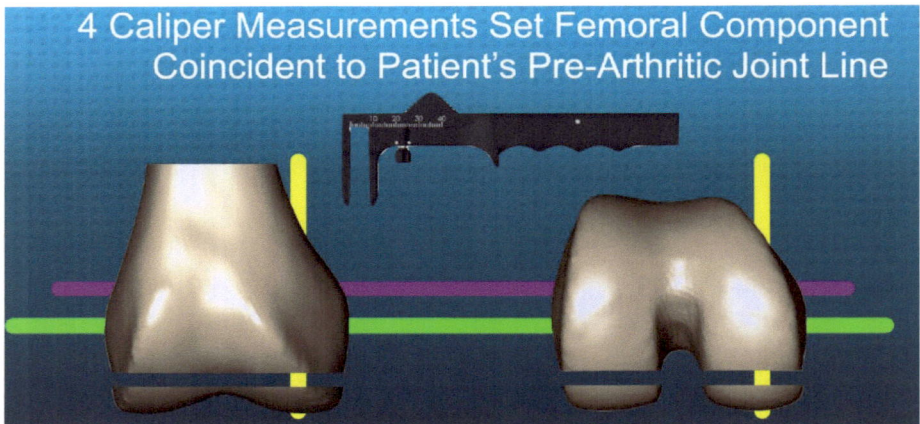

4 Caliper Measurements Set Femoral Component Coincident to Patient's Pre-Arthritic Joint Line

Surgeons that use manual instruments, patient-specific instruments, navigational and robotic instruments should consider the use of the caliper to measure the bone resections to verify the placement of the patient's femoral and tibial components restore the patient's pre-arthritic joint lines.

A schematic shows a lateral pose of the distal femur and the caliper measurement of the posterior lateral femoral resection.

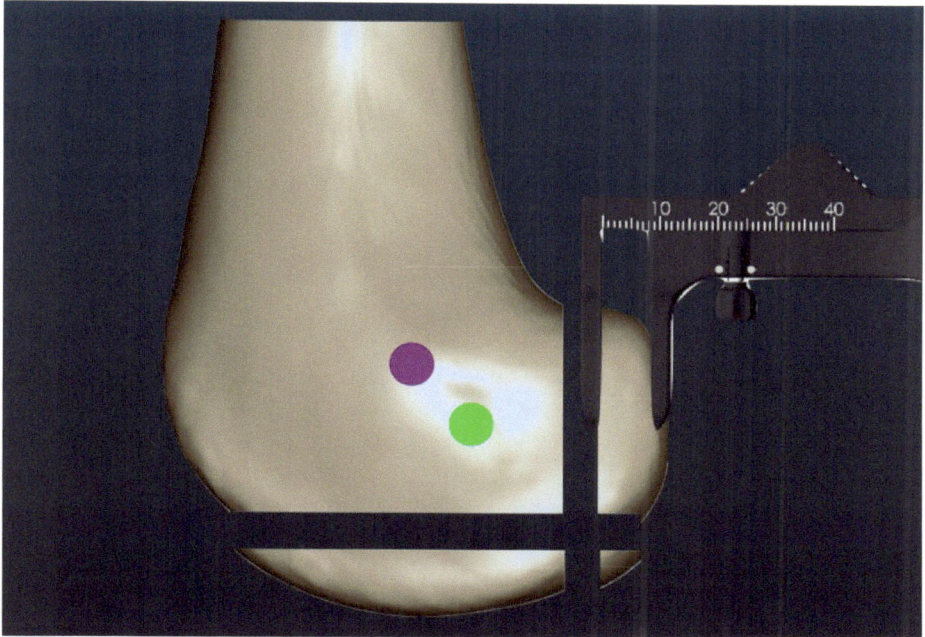

The thicknesses of the distal resections determine the proximal-distal and varus-valgus position of the femoral component. The thicknesses of the posterior resections determine the anterior-posterior and internal-external rotation position of the femoral component. So, these four resection measurements determine four of the six-degrees-of-freedom of femoral component placement, hence their importance.

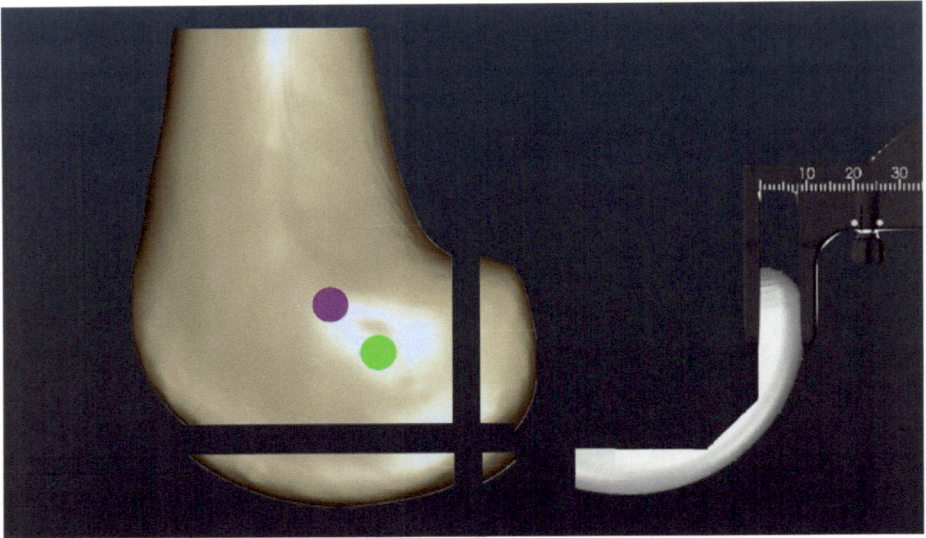

The thickness of the distal and posterior resections should match the femoral component's distal and posterior condyles after compensating 2 mm for complete cartilage loss and 1 mm for the kerf the thickness of the saw blade, which is about 1 millimeter. These intra-operative checks verify accurate execution of the surgery and restoration of the pre-arthritic femoral joint lines.

Why are the caliper verification checks so necessary? They confirm that the femoral component is kinematically aligned and that the components' axes are co-aligned with those of the native knee.

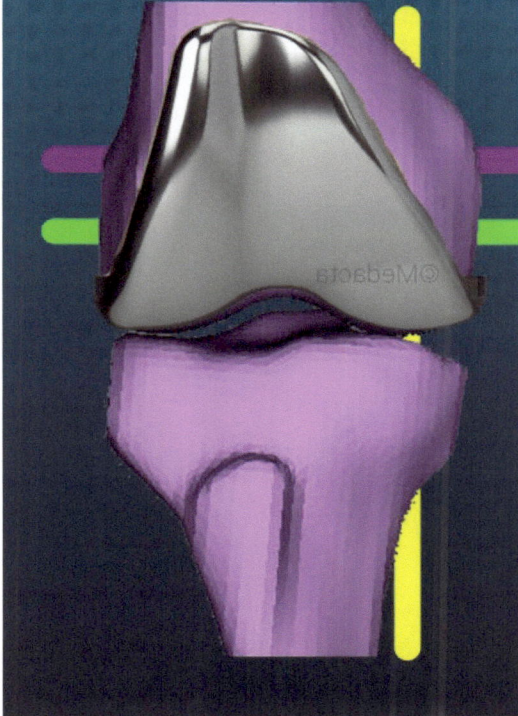

KA Sets Femoral
Component to
Restore Pre-Arthritic
Distal Joint
Line and Axes

Mechanical alignment sets the femoral component varus to the patient's pre-arthritic distal femoral joint line in 84% of patients.

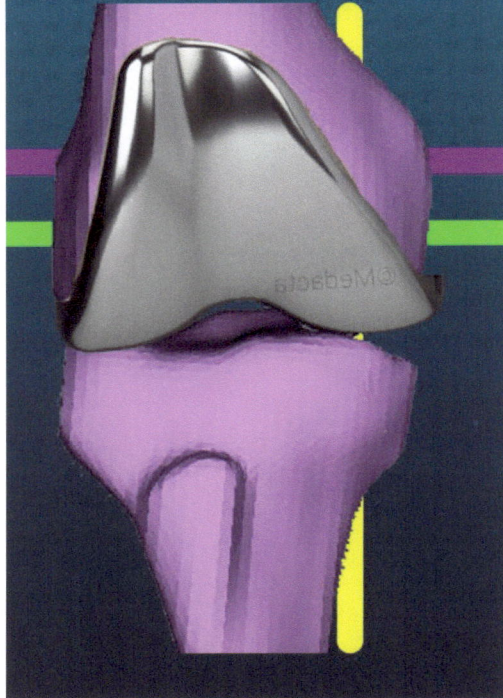

MA Sets Femoral Component in Varus to Joint Line and Deviates From Axes

When mechanical alignment sets the femoral component in varus relative to the pre-arthritic distal joint line in the coronal plane, the axes deviation between the components and knee cause ligamentous tightening slackening, which shouldn't occur. Axes deviation leads to motion loss and instability. Reducing the risk of motion loss requires the release of healthy ligaments, making no sense when kinematic alignment retains healthy ligaments.

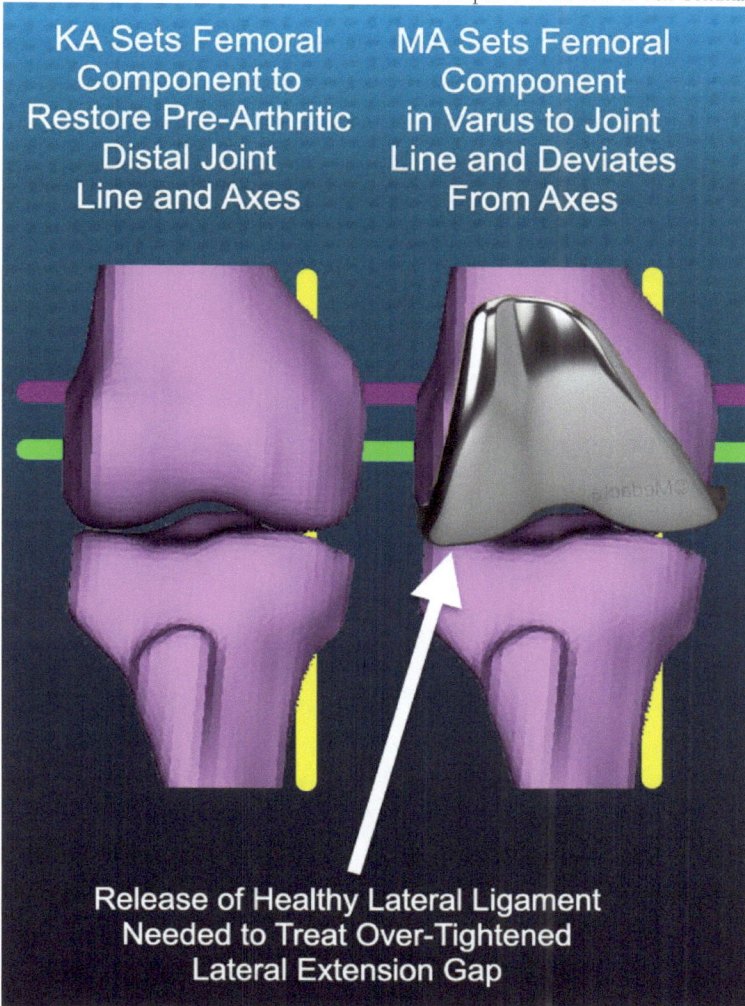

KA Sets Femoral Component to Restore Pre-Arthritic Distal Joint Line and Axes

MA Sets Femoral Component in Varus to Joint Line and Deviates From Axes

Release of Healthy Lateral Ligament Needed to Treat Over-Tightened Lateral Extension Gap

Mechanical alignment causes similar problems in the axial plane because the femoral component is not coincident with the patient's posterior pre-arthritic joint line. An external rotation deviation of the femoral component tightens the posterior lateral compartment requiring the release of a healthy lateral collateral ligament to compensate for the impending stiffness.

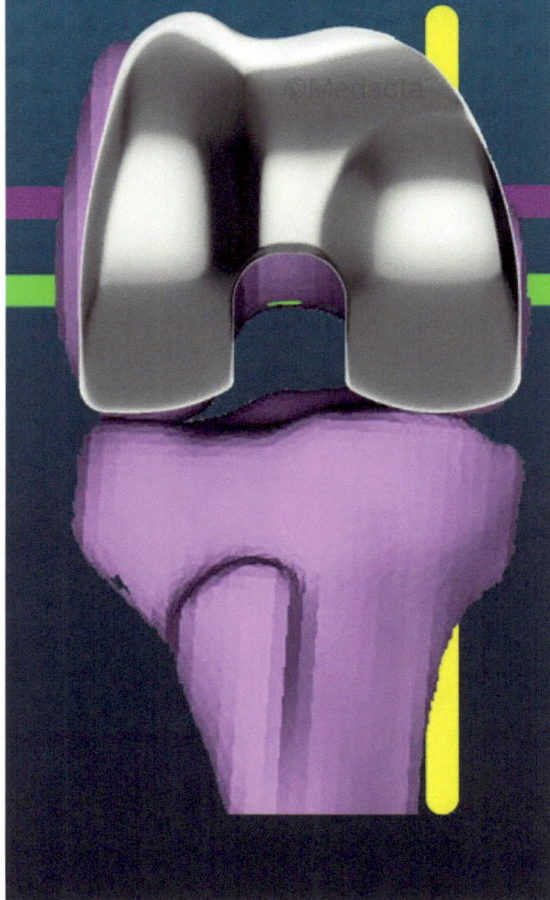

KA Sets Femoral
Component to
Restore Pre-Arthritic
Posterior Joint
Line and Axes

Once again, kinematic alignment solves this problem by co-aligning the axes of the femoral component with those of the knee.

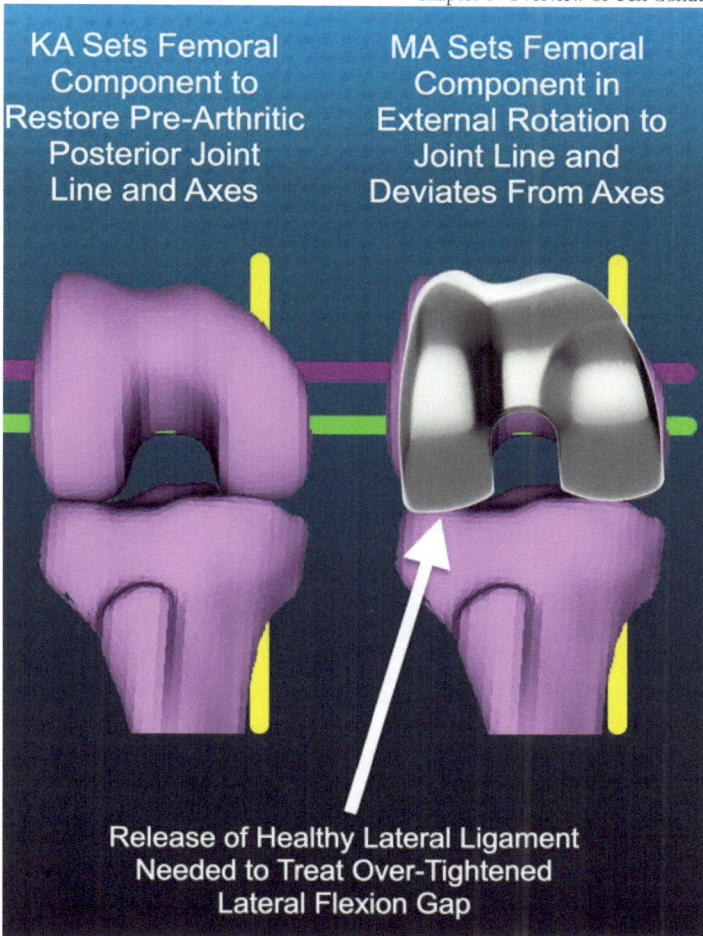

KA Sets Femoral Component to Restore Pre-Arthritic Posterior Joint Line and Axes

MA Sets Femoral Component in External Rotation to Joint Line and Deviates From Axes

Release of Healthy Lateral Ligament Needed to Treat Over-Tightened Lateral Flexion Gap

The topics of the 10 Commandments clear-up several misconceptions about calipered kinematic alignment and provide strategies for reducing the risk of stiffness, patellofemoral instability, tibiofemoral instability, and tibial component failure.

The TEN COMMANDMENTS of Calipered Kinematically Aligned TKA

1. Thou shalt not perform kinematic alignment without recording caliper measurements of the bone resections, filling out a verification sheet, and following a decision-tree.

2. Thou shalt not leave a component proud or recessed from the pre-arthritic joint surface as the knee will feel tight or loose to the patient.

3. Thou shalt not make a straight leg instead straighten the leg to the pre-arthritic alignment.

4. Thou shalt not worry about varus failure of the tibial component as this is caused by mechanical alignment and alleviated by kinematic alignment.

5. Thou shalt not profess allegiance to the center of the femoral head and ankle.

6. Thou shalt not deny treatment to knees with severe varus, valgus, and flexion deformities

7. Thou shalt not commit to the first tibial cut instead use a recut guide to optimize varus-valgus stability in extension and restore native laxity in flexion.

8. Thou shalt not flex the femoral component more than a couple of degrees or risk late-onset patellofemoral instability

9. Thou shalt not deviate from the patient's pre-arthritic slope more than a few degrees when using a CR insert or risk early-onset tibial component failure from posterior overload

10. Thou shalt not intentionally remove the PCL, though when cut, reduce the posterior tibial slope to compensate for the increase in laxity in the flexion space.

So why follow the 10 Commandments of Calipered Kinematically Aligned TKA?

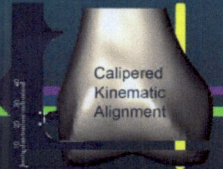

Why Follow the Ten Commandments of Calipered KA TKA?

- Fills your professional life with happy patients, busy OR, leisurely office, and few complications
- Provides principles for learning, improving outcomes, and advancing surgical technique and implant design
- Using a caliper makes you a better surgeon

Calipered Kinematic Alignment

Calipered kinematic alignment will fill your professional life with happier patients, a busy operating room, and a leisurely office with few complications and patient complaints. You can help more patients and do more surgery without exhausting yourself.

Those that use a caliper will become a much better surgeon by identifying when the resection misses the target, thereby enabling correction before implanting the components.

We hope you enjoyed the overview-now let's get to the 10 Commandments:

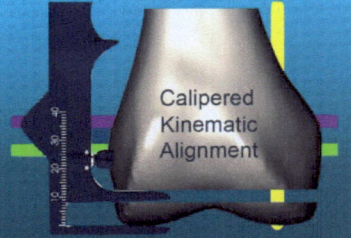

OVERVIEW:

DEFINE CALIPERED KINEMATIC ALIGNMENT!

WHY FOLLOW THE TEN COMMANDMENTS?

Chapter 1 The 1st Commandment

The 1st Commandment is **Thou Shalt Not Perform Kinematic Alignment Without Recording Caliper Measurements of the Bone Resections, Filling Out a Verification Sheet, and Following a Decision Tree**. These are the three components of the calipered kinematic alignment technique.

1.   THOU SHALT NOT PERFORM KINEMATIC ALIGNMENT WITHOUT RECORDING CALIPER MEASUREMENTS OF THE BONE RESECTIONS, FILLING OUT A VERIFICATION SHEET, AND FOLLOWING A DECISION-TREE.

Measuring the thicknesses of the femoral component's distal and posterior condyles with the caliper determines the target thickness for the distal and posterior femoral resections. Consider a femoral component with a distal thickness of 9 mm and a posterior thickness of 8 mm.

The target for the distal and posterior femoral resections should equal these thicknesses after compensating ~ 1 millimeter for the kerf of the saw blade and ~ 2 millimeters when there is full-thickness cartilage wear. So, when there is no cartilage, the distal femoral resections' thickness should equal 8 millimeters, and the posterior resections should equal 7 mm. When there is full-thickness cartilage wear, these measurements will be 2 mm thinner.

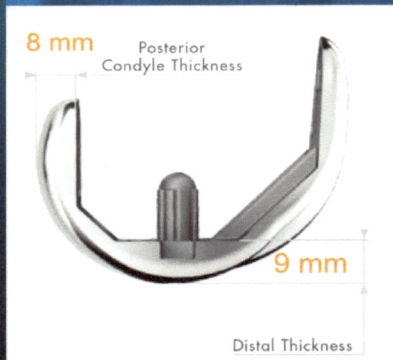

Target Thickness for Distal and Posterior Femoral Resections

- Compensate 1 mm for the kerf of the saw blade and 2 mm for cartilage wear
  - Distal resections without wear should equal 8 mm
  - Posterior resections without wear should equal 7 mm

GⱭⱭK SPHERE
MEDIALLY STABILIZED KNEE

8 mm   Posterior Condyle Thickness

9 mm

Distal Thickness

Let's use a right knee with medial wear or a varus deformity, cut the distal femur, and measure the resection thickness with the caliper.

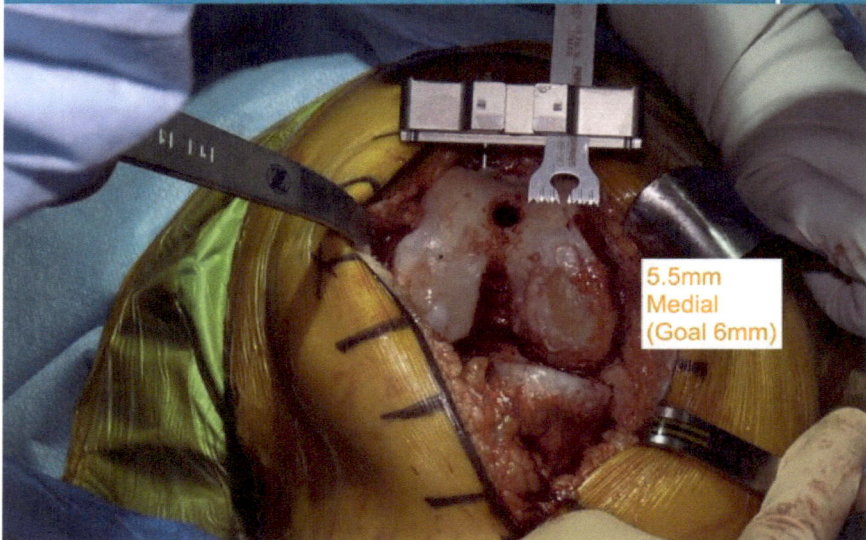

Cut Distal Femur and Measure Thickness of Resections with Caliper

5.5mm Medial (Goal 6mm)

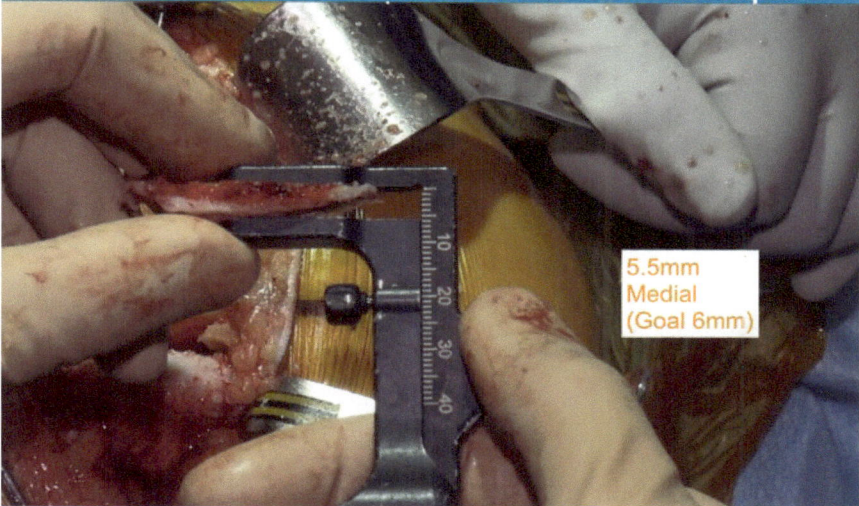

Cut Distal Femur and Measure
Thickness of Resections with Caliper

5.5mm
Medial
(Goal 6mm)

Because cartilage is missing from the distal medial femur, the resection should measure 6 mm in thickness. However, it was 5.5 mm. The distal lateral femoral was unworn, and the resection of 8 mm (+ 1 mm for the kerf of the blade) matches the target.

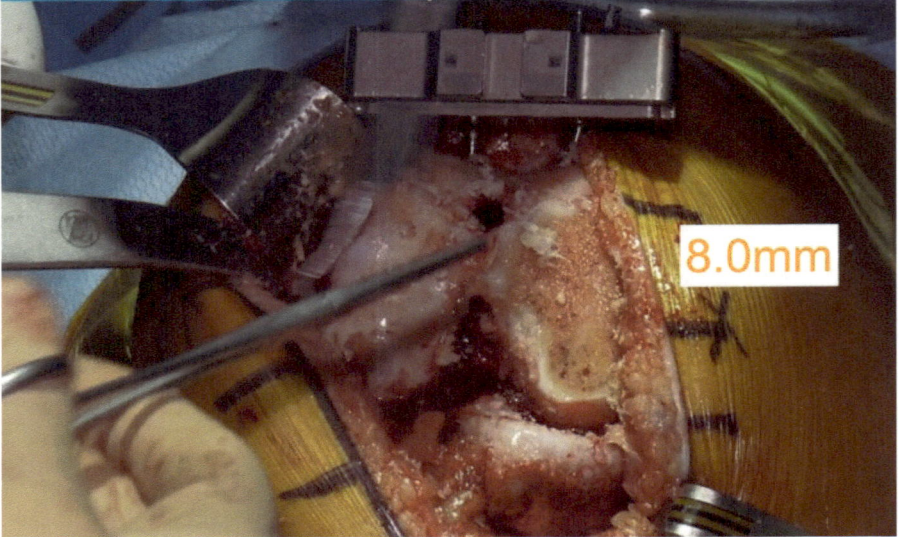

Cut Distal Femur and Measure
Thickness of Resections with Caliper

8.0mm

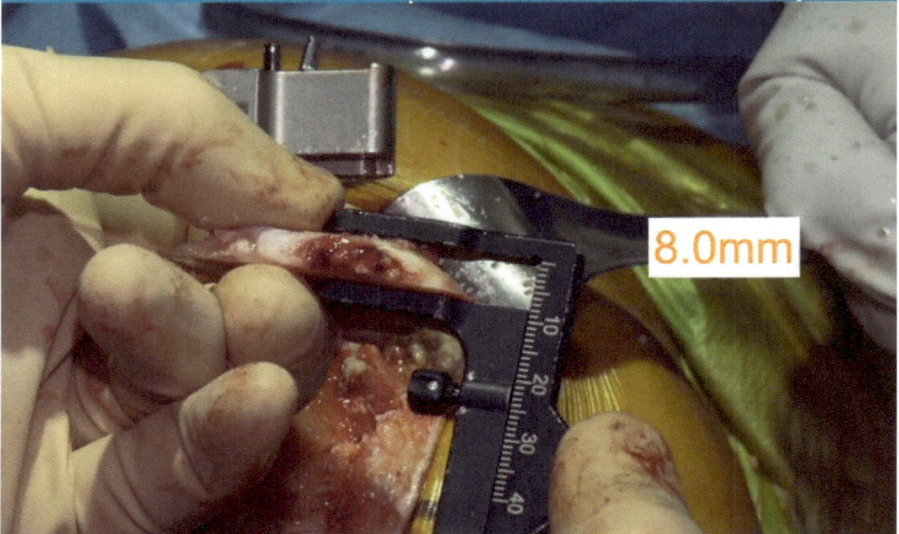

Cut Distal Femur and Measure
Thickness of Resections with Caliper

8.0mm

So, there is a 0.5 mm under-resection on the distal medial femoral resection. The saw removed another 0.5 mm of distal medial bone using the 1.2 mm thickness of the saw blade functioned as a depth indicator, which corrected a 0.5 degree valgus deviation of the femoral component from the patient's pre-arthritic distal femoral joint line.

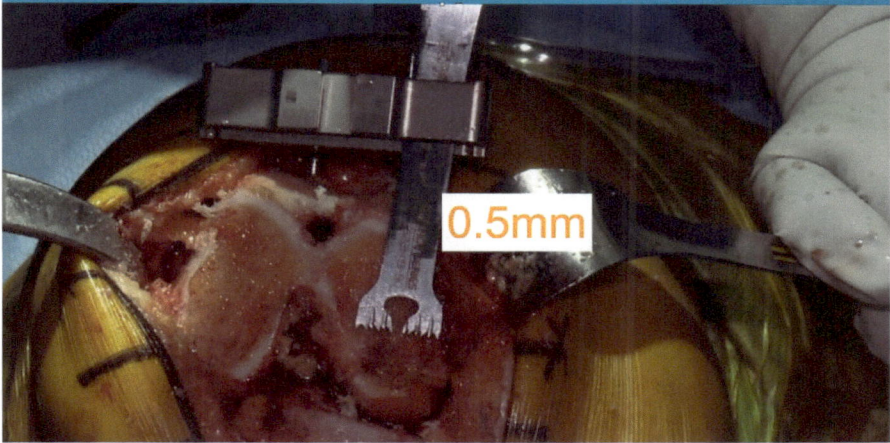

Match Target by Recutting 0.5 mm from Distal Medial Condyle and Record Thickness

0.5mm

The next step is to record the thickness of the distal femoral resections on the verification sheet, which serves as a 'report card' of the components' setting in calipered kinematically aligned TKA.

Medacta
International

RECORD OF VERIFICATION CHECKS FOR CALIPERED KINEMATICALLY ALIGNED MEDACTA GMK SPHERE TKA

RESET

SURGEON   PATIENT CODE   DATE (DD/MM/YYYY)

KNEE ☐ RIGHT ☐ LEFT

OA DEFORMITY ☐ VARUS ☐ VALGUS ☐ PF

A/P OFFSET

EXPOSURE   mm   TRIALING   mm   DIFFERENCE   mm

ACL CONDITION

☐ INTACT   ☐ TORN   ☐ GRAFT

DISTAL FEMORAL RESECTION

Target Thickness: 8mm Unworn, 6mm Worn (No Cartilage)
When initial thickness misses target - recut or use a washer

MEDIAL CONDYLE			LATERAL CONDYLE		
☐ UNWORN		☐ WORN	☐ UNWORN		☐ WORN
INITIAL THICKNESS		mm	INITIAL THICKNESS		mm
RECUT	☒ N ☒ Y	mm	RECUT	☒ N ☒ Y	mm
WASHER	☒ N ☒ Y	mm	WASHER	☒ N ☒ Y	mm
FINAL THICKNESS	0.0	mm	FINAL THICKNESS	0.0	mm

POSTERIOR FEMORAL RESECTION

Target Thickness: 7mm Unworn, 5mm Worn (No Cartilage)
When initial thickness misses target - recut

MEDIAL CONDYLE			LATERAL CONDYLE		
☐ UNWORN		☐ WORN	☐ UNWORN		☐ WORN
INITIAL THICKNESS		mm	INITIAL THICKNESS		mm
RECUT	☒ N ☒ Y	mm	RECUT	☒ N ☒ Y	mm
FINAL THICKNESS	0.0	mm	FINAL THICKNESS	0.0	mm

TIBIAL RESECTION

Target: Equal Thickness Measured at Base of Tibial Spines

☐ MEDIAL   ☐ LATERAL

☐ MEDIAL   ☐ LATERAL

mm   mm

PCL CONDITION

☐ INTACT   ☐ TORN   ☐ EXCISED

TIBIAL V-V RECUT   deg

TIBIAL SLOPE RECUT   deg

FINAL CHECK WITH SPACER BLOCK AND TRIAL COMPONENTS

NEGLIGIBLE V-V LAXITY IN EXTENSION

2-3 MM OF LATERAL OPENING WITH VARUS LOAD IN 15-30° OF FLEXION

G⏢K SPHERE
MEDIALLY STABILIZED KNEE

FEMUR SIZE	TIBIA SIZE	INSERT THICKNESS	PATELLA SIZE
		☐ CR	
		☐ CS	

© 2018 Medacta International SA. All rights reserved   99 RMPAA 15 REV.00

DISTAL FEMORAL RESECTION
Target Thickness: 8mm Unworn, 6mm Worn (No Cartilage)
When initial thickness misses target - recut or use a washer

MEDIAL CONDYLE		
☐ UNWORN	X	WORN
INITIAL THICKNESS	**5.5** mm	
RECUT N X	**0.5** mm	
WASHER N Y	mm	
FINAL THICKNESS	**6** mm	

LATERAL CONDYLE		
X UNWORN	☐	WORN
INITIAL THICKNESS	**8** mm	
RECUT X Y	mm	
WASHER N Y	mm	
FINAL THICKNESS	**8** mm	

The expanded view of the intraoperative verification sheet shows the cuts were 'kinematically correct'. The initial distal medial femoral resection was 5.5 mm. The re-cut removed 0.5 mm more, making a 6 mm resection that matched the kinematic alignment target. The lateral condyle was unworn, so the distal lateral resection's 8 mm thickness matched the kinematic alignment target.

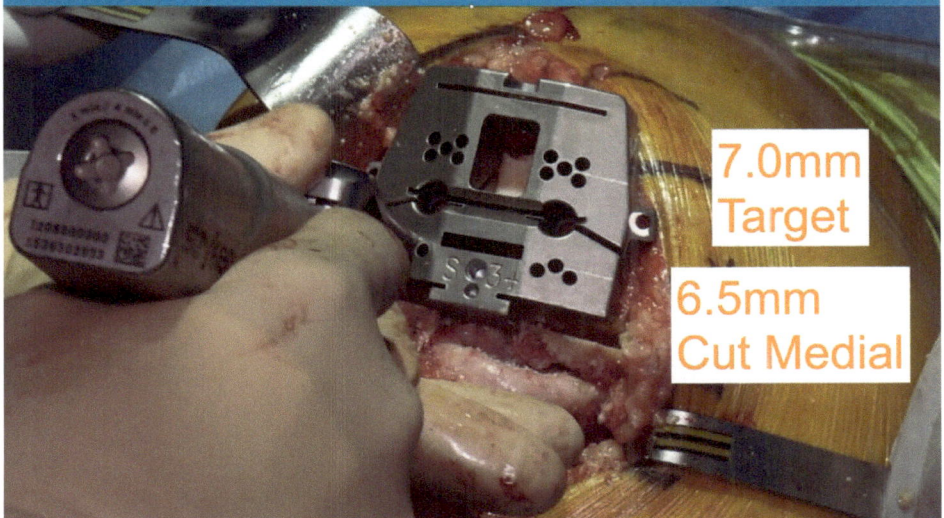

Cut Posterior Femur and Measure Thickness of Resections with Caliper

7.0mm Target

6.5mm Cut Medial

Make the posterior cuts before making the anterior and chamfer cut, enabling anterior-posterior and internal-external rotational adjustments of the 4-in-1 block when the posterior resections don't match the target. For a femoral component with an 8 mm thick posterior condyle, the target resection is 7 mm (+ 1 mm for the kerf of the blade) when there is no cartilage loss.

POSTERIOR FEMORAL RESECTION
Target Thickness: 7mm Unworn, 5mm Worn (No Cartilage)
When initial thickness misses target - recut

MEDIAL CONDYLE		
X UNWORN	☐ WORN	
INITIAL THICKNESS	**6.5** mm	
RECUT	N X	**0.5** mm
FINAL THICKNESS	**7** mm	

LATERAL CONDYLE		
X UNWORN	☐ WORN	
INITIAL THICKNESS	**7** mm	
RECUT	X Y	mm
FINAL THICKNESS	**7** mm	

The expanded view of the verification sheet shows the posterior femoral resections were 'kinematically correct'. The posteromedial resection of 6.5 mm was under-resected by 0.5 mm. Palpation indicated that the saw blade skived, so the posterior medial femur was fine-tuned by removing another 0.5 mm of bone. Because the posterior lateral condyle was unworn, the 7 mm thickness of the resection matched the target. The recording of the distal and posterior femoral resections' thicknesses verified the proximal-distal, anterior-posterior positions, and the varus-valgus and internal-external rotational orientations of the was set coincident with the patient's pre-arthritic joint line.

A visual check of the tibial resection's posterior slope determines whether the patient's pre-arthritic slope is restored, which is a requirement with retention of the posterior cruciate ligament.

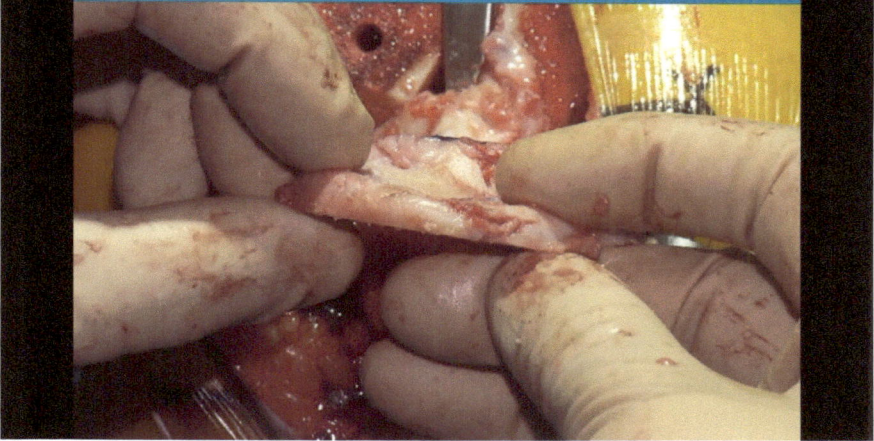

Check Slope and Measure Thickness of Medial and Lateral Tibial Condyles with Caliper

Fine-tune the posterior slope when less than pre-arthritic to reduce correct a tight flexion space. Fine-tune the posterior slope when greater than pre-arthritic to correct a loose flexion space.

Measure the thickness of the medial and lateral compartments of the tibial resection with the caliper at the base of the tibial spine away from areas of wear.

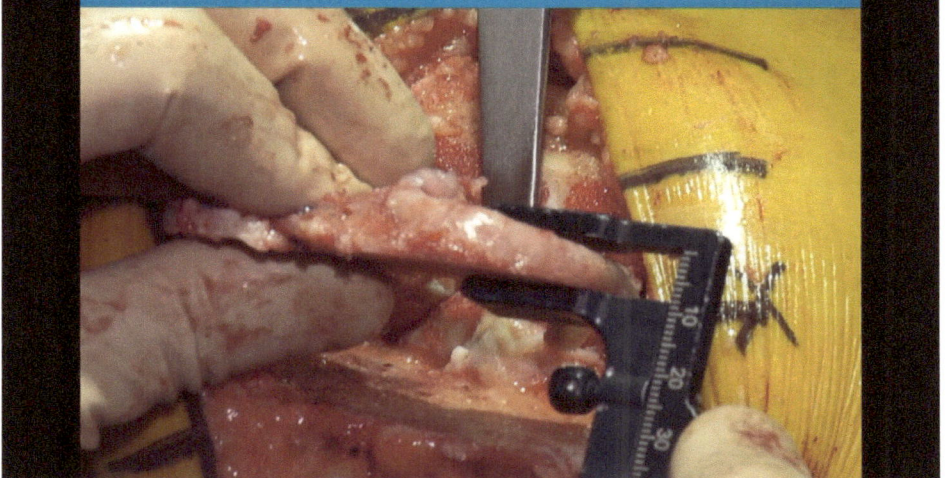

Check Slope and Measure Thickness of Medial and Lateral Tibial Condyles with Caliper

The measurements are made at the base of the tibial spines where cartilage is typically intact and recorded on the verification sheet.

When the medial compartment is thinner than the lateral compartment, adjust the medial cut until there is negative varus-valgus laxity in full extension with a spacer block and trial components. When the lateral compartment is thinner than the medial compartment, adjust the lateral cut until there is negative varus-valgus laxity in full extension with a spacer block and trial components.

Include the verification sheet values in the operative note to record that the surgeon kinematically aligned the femoral component.

Consider following the 1st Commandment:

1.   THOU SHALT NOT PERFORM KINEMATIC ALIGNMENT WITHOUT RECORDING CALIPER MEASUREMENTS OF THE BONE RESECTIONS, FILLING OUT A VERIFICATION SHEET, AND FOLLOWING A DECISION-TREE.

Chapter 2 The 2nd Commandment

The 2nd Commandment is **Thou Shalt Not Leave a Component Proud or Recessed from the Pre-Arthritic Joint Surface as the Knee Could Feel Tight or Loose to The Patient**. Changing the patient's pre arthritic joint lines malaligns the components to the three kinematic axes.

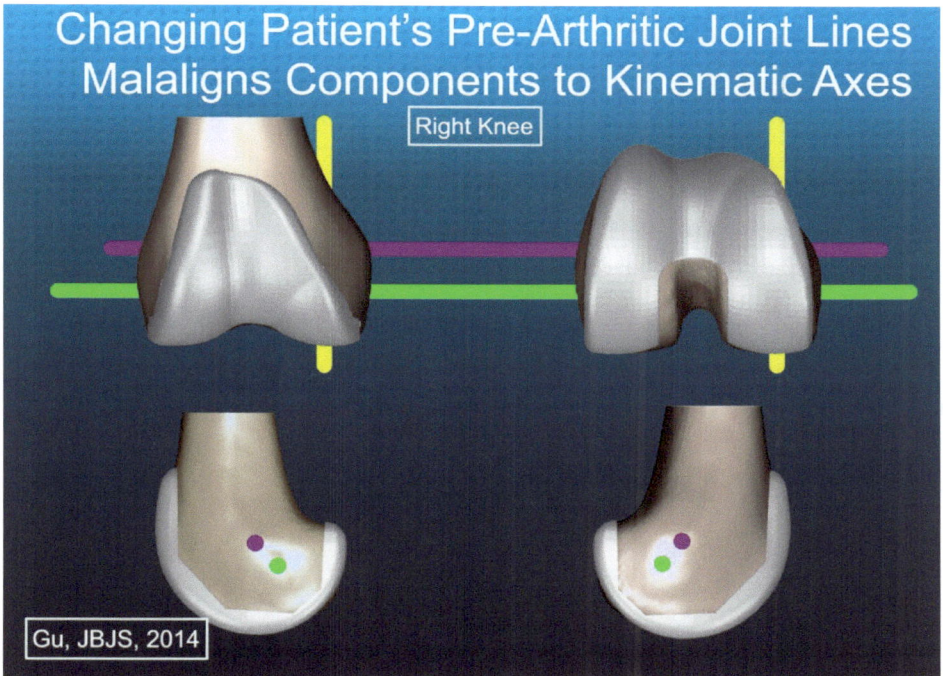

A schematic of the right knee shows the green and magenta line, which are the femur's transverse axes about which the tibia and patella flex and extend. The yellow vertical line is the axes about which the tibia internally and externally rotates on the femur and is centered medial and lateral and slightly posterior in the medial compartment. A femoral component set in varus, which happens in 84% of limbs with mechanical alignment, tightens the distal lateral gap and slackens the distal medial gap.

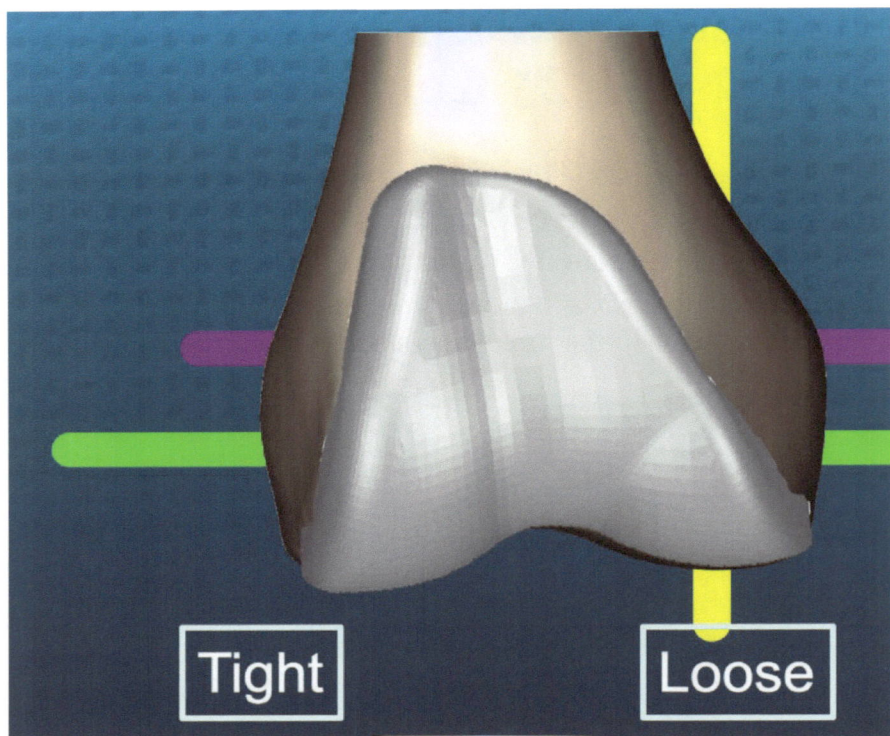

Translation of the femoral component posteriorly tightens the posterior medial and lateral gaps.

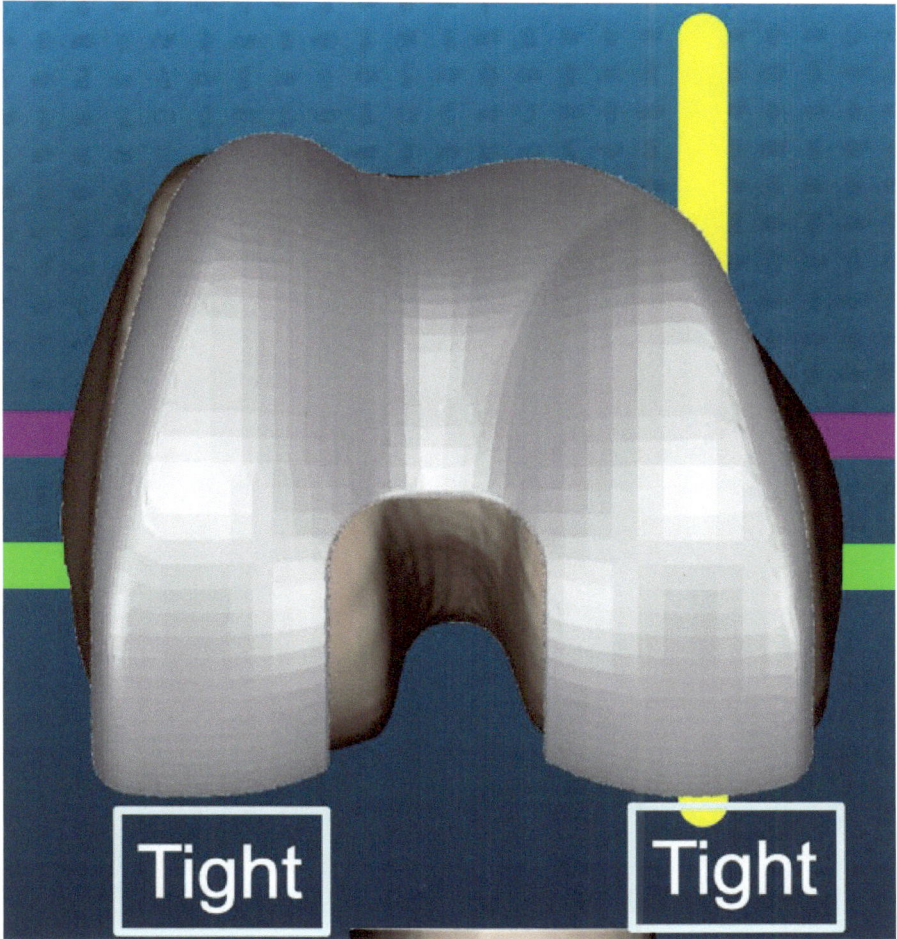

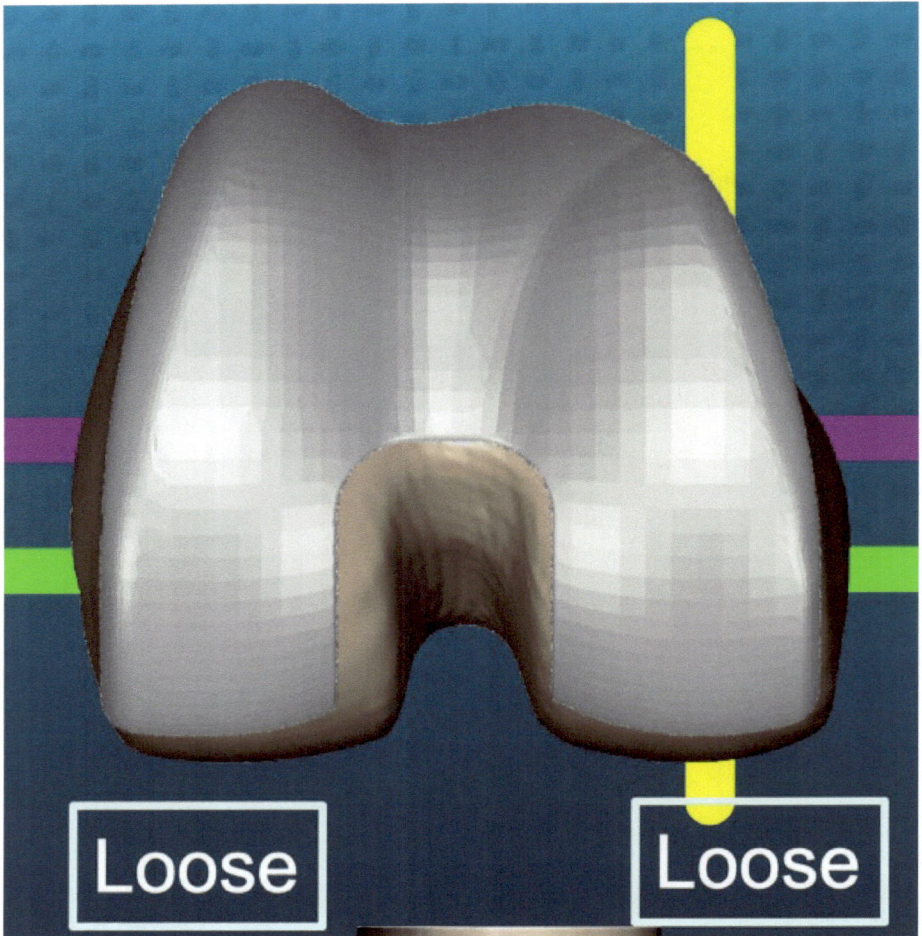

Translation of the femoral component anterior loosens the posterior medial and lateral gaps. Adding rotation creates a myriad of complexities that leave the TKA unbalanced.

Let's examine what happens when we translate the femoral component proximal and posterior on the lateral femoral condyle.

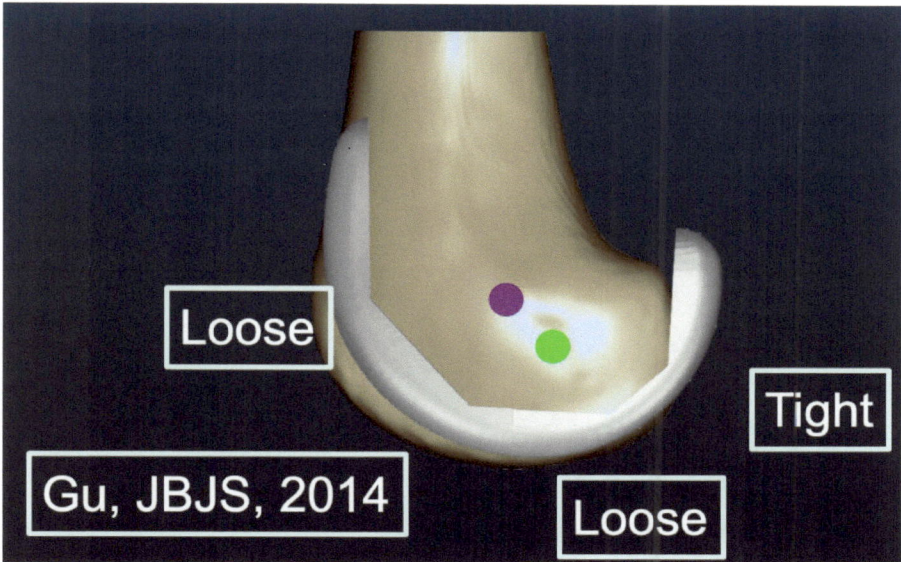

The lateral view shows the adverse consequence of patella-femoral tracking because the lateral retinacular ligament is loose, the distal gap is loose, and the posterior gap is tight. A healthy ligament release cannot reconcile the difference between the loose extension and tight flexion gap in the lateral compartment. Only kinematic alignment of the femoral component reconciles the loose-tight extension-flexion imbalance in a compartment.

Let's look at the medial side and the adverse consequences from a distal and anterior translation of the femoral component, opposite the one on the lateral side.

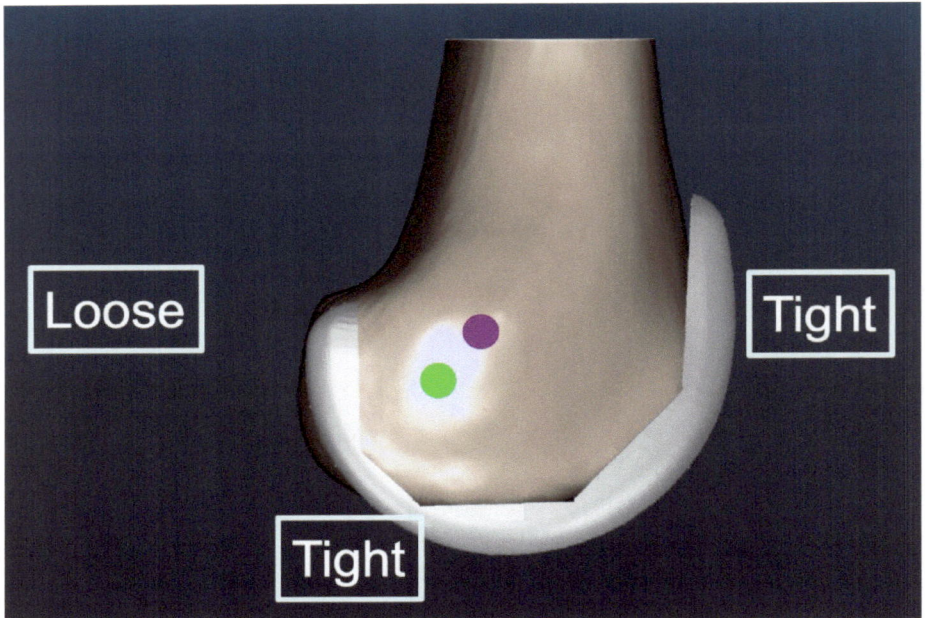

The medial view shows the lateral retinacular ligament is loose, the distal gap is tight, and the posterior gap is loose. A healthy ligament release cannot reconcile the difference between the tight extension and loose flexion gap. Only repositioning the femoral component reconciles the tight-loose extension-flexion imbalance in a compartment.

A thought experiment that uses an outside patio door as the femoral component and the frame as the ligaments and tibia helps understand the adverse consequences of not using kinematic alignment.

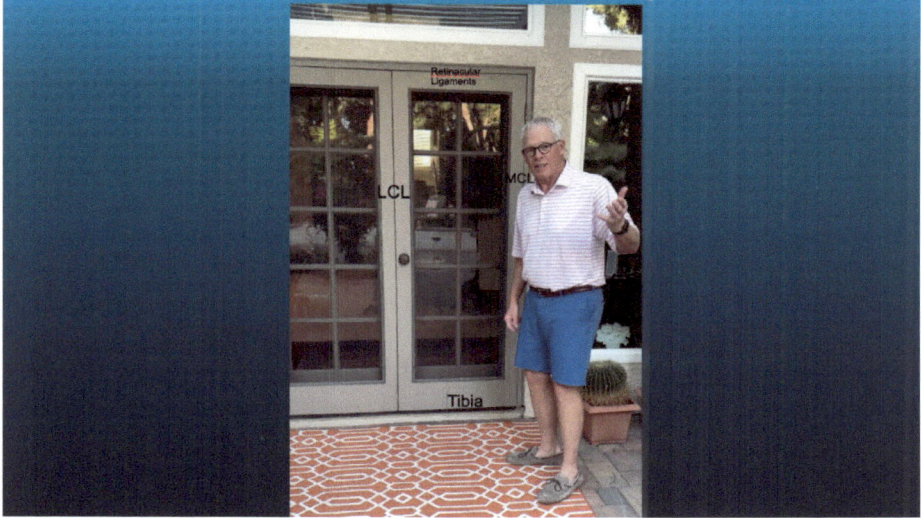

Back Yard Analogy Between Setting the Femoral Component and Door Correctly in the Frame

Let's assume that the bottom of the door frame is the tibia, the right side is the medial collateral ligament (MCL), the left side is the lateral collateral ligament (LCL), and the top is the retinacular ligaments.

When both the screen door facing the outside of the house and the femoral component is 'kinematically aligned', there is smooth motion of the door and knee without binding, and the fit is tight.

Inside, the door that secures the house and blocks intruders is hung like a typical mechanically aligned femoral component as it doesn't fit the frame, and the ligaments are either too tight or too loose.

Back Yard Analogy Between Setting the Femoral Component and Door Correctly in the Frame

When I try to open this door, it hangs up, and my wife's been wanting me to fix it. If we come closer, you can see what's happening.

As a door set crooked in the frame 'hangs- up' when opening and closing, a mechanically aligned femoral component set crooked to the patient's pre-arthritic joint lines results in a knee that is stiff and slow to regain motion. Since kinematic alignment restores the patient's pre-arthritic joint line, recovery is fast and less painful.

The following schematics of a door reinforce what we just learned!

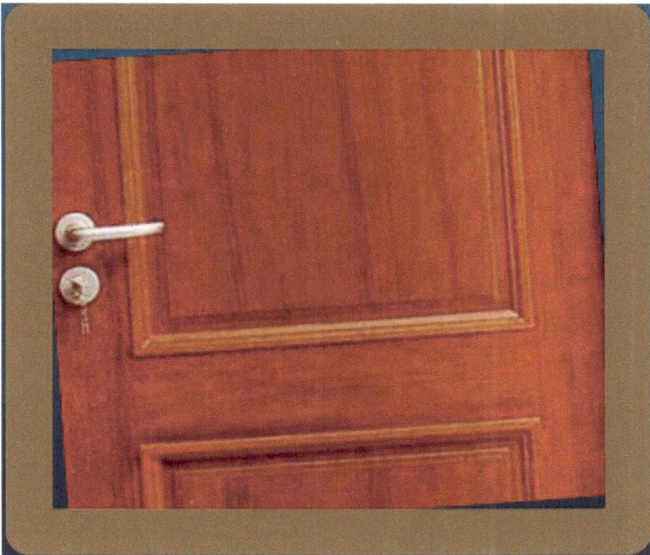

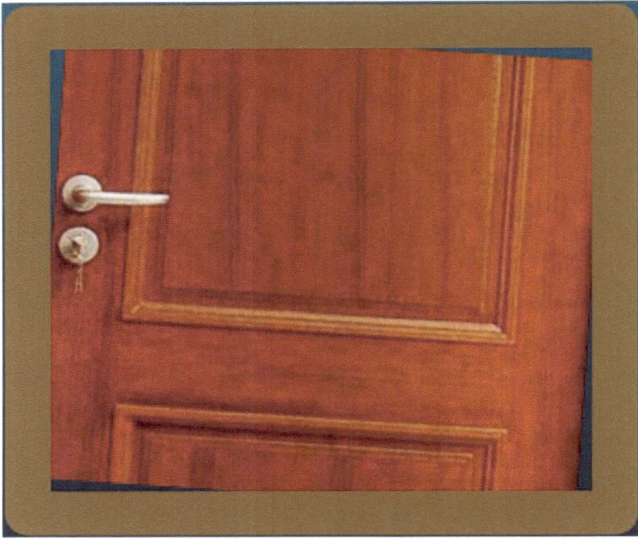
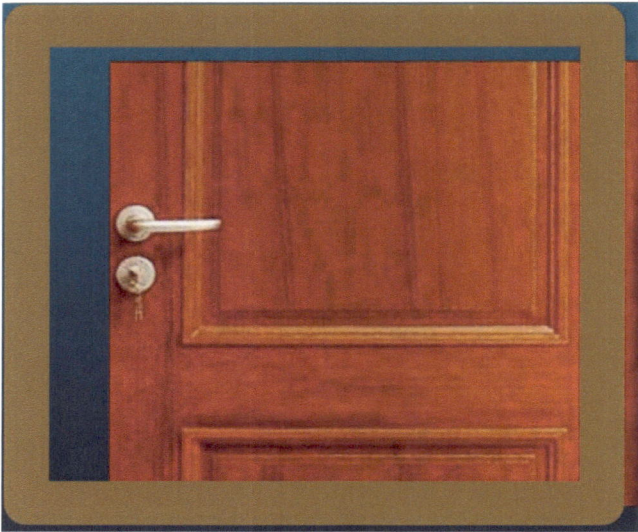

The carpenter has many ways to hang the door incorrectly; they can rotate it and translate it relative to the frame.

The following schematics of a femoral component reinforce what we just learned!

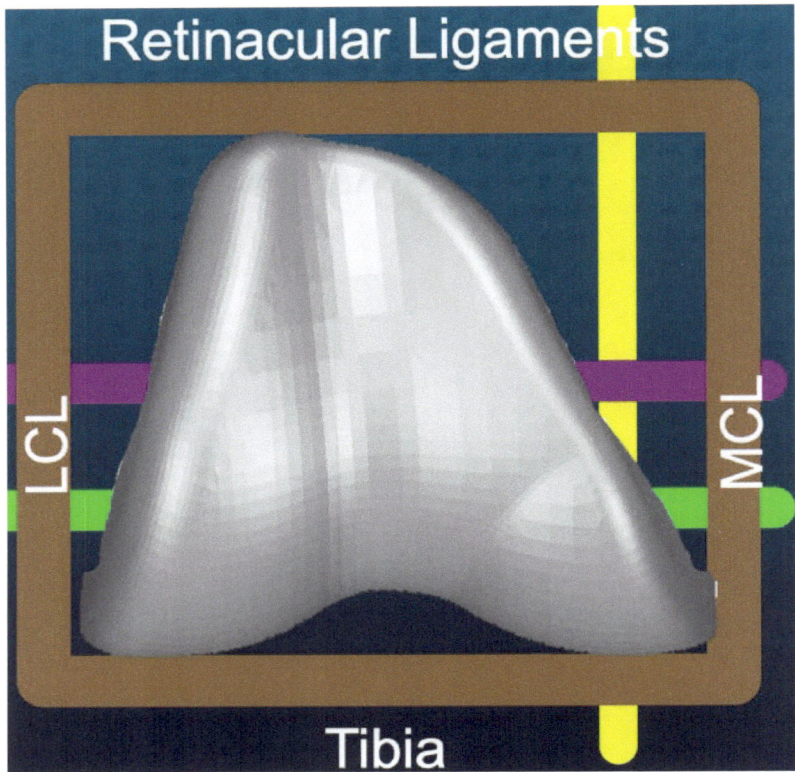

In the coronal plane, the co-alignment of the femoral component's axes with the three kinematic axes of the knee does not alter the frame comprised of the ligaments and tibia.

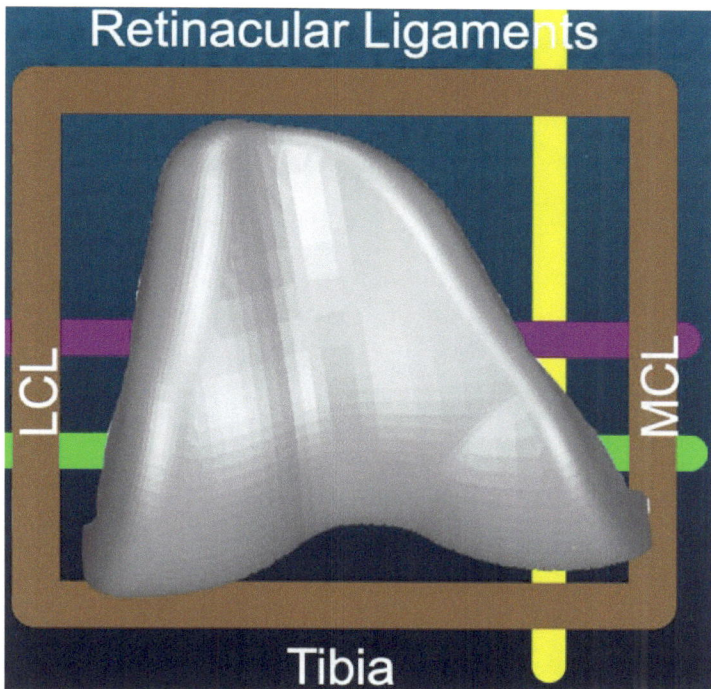

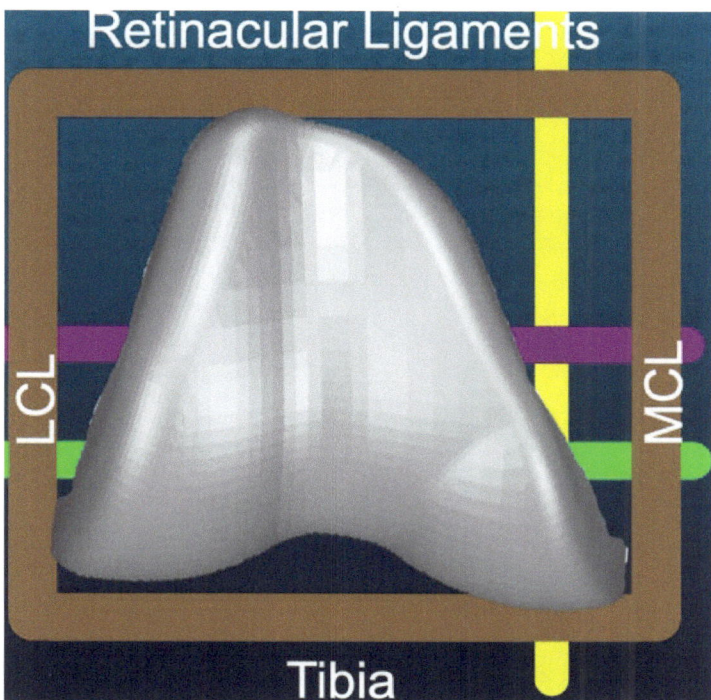

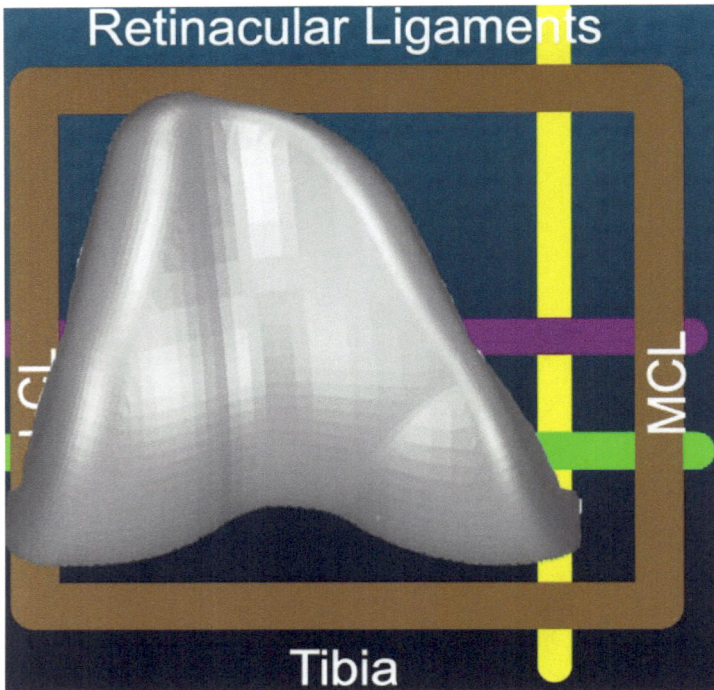

In the coronal plane, the surgeon has many ways to hang the femoral component incorrectly; they can mal-rotate it too varus-valgus, too flexed-extended, and can mal-translate it too proximal-distal relative to the frame.

In the axial plane, the co-alignment of the femoral component's axes with the three kinematic axes of the knee does not alter the frame comprised of the ligaments and tibia.

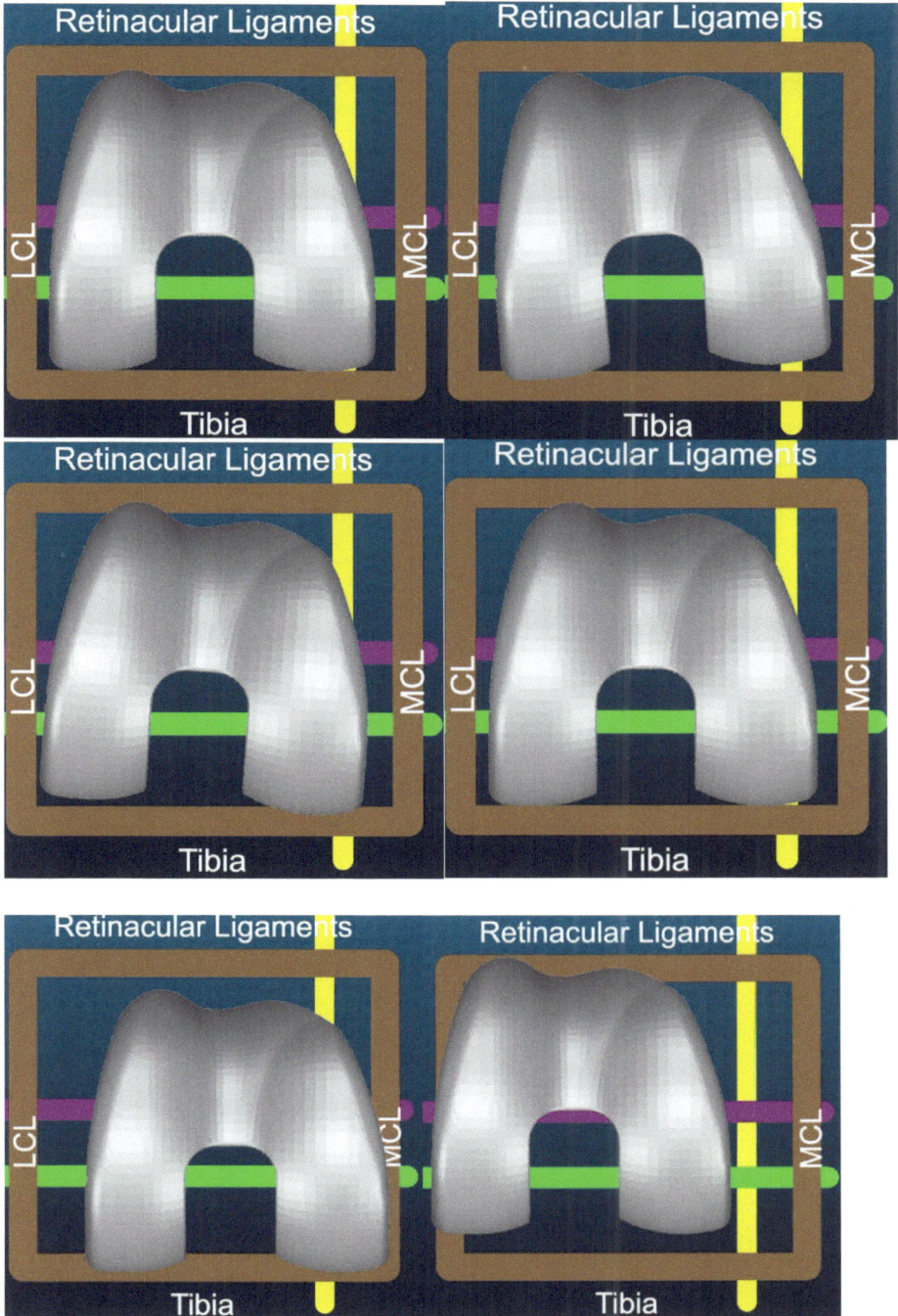

In the axial plane, the surgeon has many ways to hang the femoral component incorrectly; they can mal-rotate it too internal-external and mal-translate it too anterior-posterior medial-lateral relative to the frame comprised of the healthy collateral, retinacular, and posterior cruciate ligaments.

Consider following the 2nd Commandment:

2.   THOU SHALT NOT LEAVE A COMPONENT PROUD OR RECESSED FROM THE PRE-ARTHRITIC JOINT SURFACE AS THE KNEE WILL FEEL TIGHT OR LOOSE TO THE PATIENT.

Chapter 3 The 3rd Commandment

The 3rd Commandment is **Thou Shalt Not Make a Straight Leg Instead Straighten the Leg to the Patient's Pre-Arthritic Alignment.**

3.   THOU SHALT NOT MAKE A STRAIGHT LEG INSTEAD STRAIGHTEN THE LEG TO THE PATIENT'S PRE-ARTHRITIC ALIGNMENT.

Treatment of a valgus knee deformity back in 2007 helped clarify the correct post-operative target for kinematic alignment.

How Much Should We Correct the Valgus Knee?

The initial review of the post-operative scanogram (left) suggested the patient's right knee and leg alignment looked too valgus.

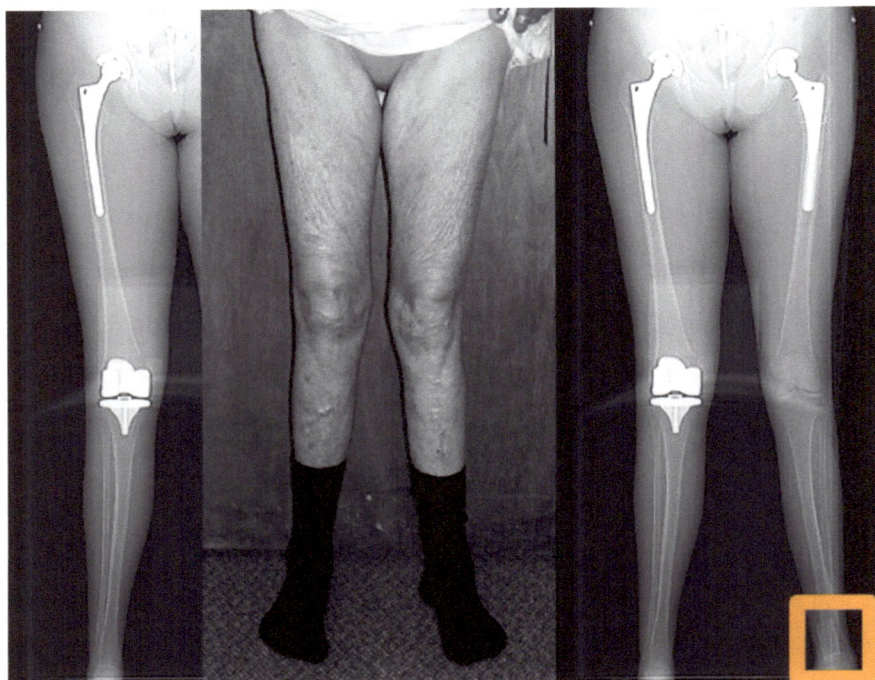

However, the photograph showing the patient standing (middle) and the post-operative scanogram (right) showing no arthritis in the left showed that kinematic alignment restored the same valgus.

This case taught that the kinematic alignment target is the restoration of the patient's pre-arthritic joint line and that the correction to this target should not be restricted. The use of mechanical alignment would have over-corrected her to neutral with the adverse consequences of not orienting the knee and ankle parallel to the floor, causing pain, wear, instability, and ankle pain.

Referencing the patient's opposite leg when it is normal is a useful surrogate for determining whether a contralateral kinematically aligned total knee arthroplasty restored the patient's pre-arthritic joint lines.

Patients Pre-Arthritic Side-to-Side Alignment is Symmetric Within $0^0 \pm 3^0$

A 2016 study by Eckhoff et al. measured the distal femoral angle, and proximal tibia angle in an extensive database of computer tomographic (CT) images of patients without lower extremity skeletal abnormalities and showed that 97% had side-to-side angles that did not differ more than 3 degrees.

Bilateral Symmetrical Comparison of Femoral and Tibial Anatomic Features

Donald G. Eckhoff, MS, MD [a], David J. Jacofsky, MD [b], Bryan D. Springer, MD [c], Michael Dunbar, MD, FRCSC, PhD [d], Jeffrey J. Cherian, DO [e], Randa K. Elmallah, MD [e], Michael A. Mont, MD [e, *], Kenneth A. Greene, MD [f]

- The DLFA of 97% of native limbs are within ± 3⁰ of contralateral side
- The PMTA of 95% of native limbs are within ± 3⁰ of contralateral side

Eckhoff, J Arthrop, 2016

The correct post-operative femoral and tibial component position after kinematic alignment is when the side-to-side difference of the distal femoral angle and proximal tibia angle is within 3 degrees, respectively.

Knowing the kinematic alignment target is a side-to-side difference helps radiographically assess the treatment of the windswept deformity with bilateral calipered kinematically aligned total knee arthroplasty.

A cruciate-retaining implant can treat both knees of most windswept deformities when performed with calipered kinematically aligned TK

Stephen M. Howell[1] · Trevor J. Shelton[2] · Manpreet Gill[3] · Maury L. Hull[1,2,4]

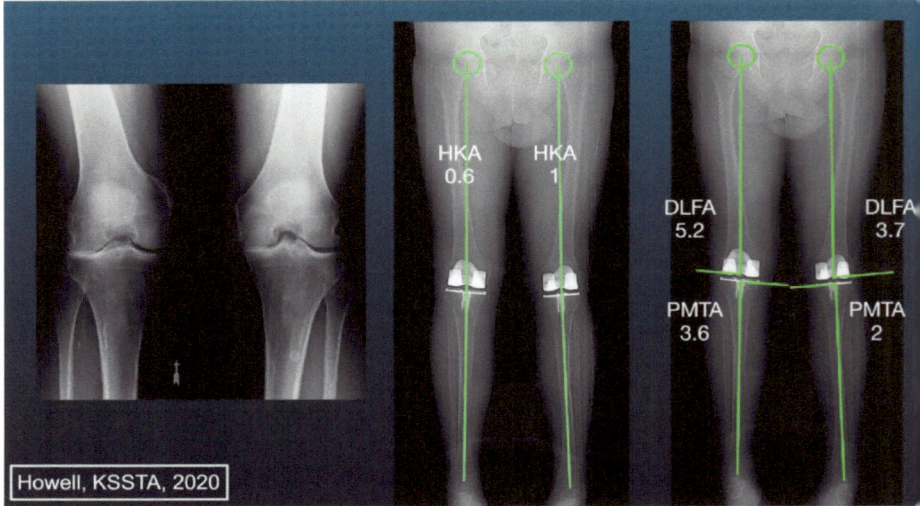

The radiographs of a typical patient with windswept deformity show valgus osteoarthritis of the right knee (knock-kneed) and varus osteoarthritis of the left knee (bow-legged) that was treated with bilateral calipered kinematically aligned TKA. Post-operatively, the left and right limbs are identically aligned. The two-degree side-to-side difference of distal lateral femoral angle (DLFA) and proximal medial tibial angle (PMTA) is within the normal side-to-side variability.

So, how accurate is calipered kinematic alignment? A 2017 study of 102 patients with a unilateral total knee arthroplasty and a normal opposite leg showed that calipered kinematic alignment restored the native left to right symmetry of the lower limb and improved function.

Does Calipered Kinematically Aligned TKA Restore Native Left to Right Symmetry of the Lower Limb and Improve Function?

Alexander J. Nedopil, MD [a, *], Avreeta K. Singh, MD [a], Stephen M. Howell, MD [b], Maury L. Hull, PhD [c]

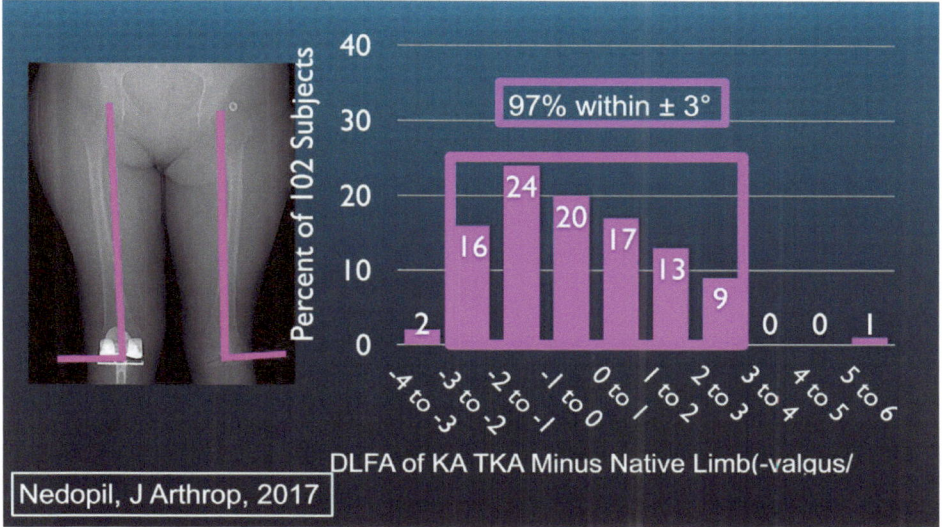

Nedopil, J Arthrop, 2017

The calipered technique restored a side-to-side difference in the distal femoral angle (DLFA) within 3 degrees in 97% of the 102 patients, which is exactly the proportion shown by Eckoff et. al. in their study of CT scans of normal subjects.

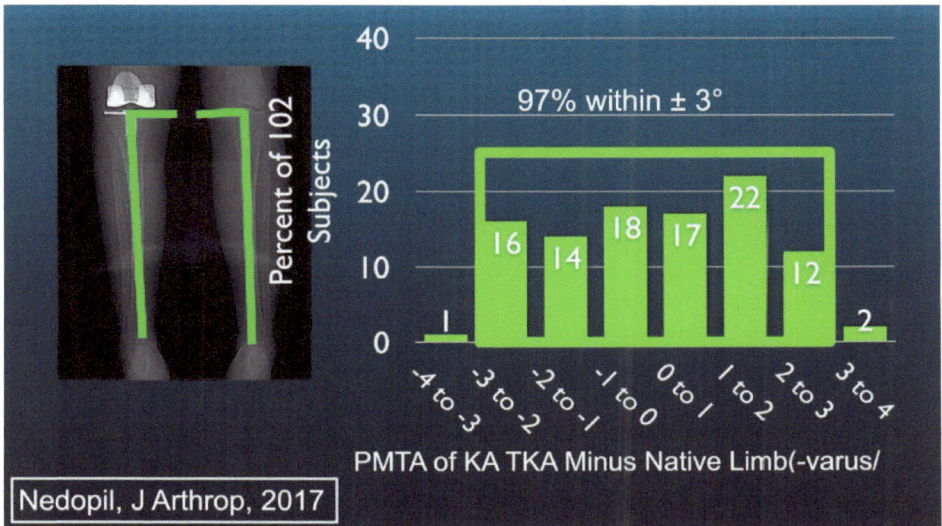

Nedopil, J Arthrop, 2017

The calipered technique restored a side-to-side difference in the proximal medial tibial angle (PMTA) within 3 degrees in 97% of the normal subjects.

Calipered KA Corrects the Osteoarthritic Deformity by Compensating for Millimeters of Joint Wear

The use of a caliper to measure the femoral bone resections and adjustment of the thickness to within ± 0.5 mm of the distal and posterior condyles of the femoral component after compensating for cartilage wear and the thickness or 'kerf' of the saw blade is highly accurate at restoring the patient's pre-arthritic joint line.

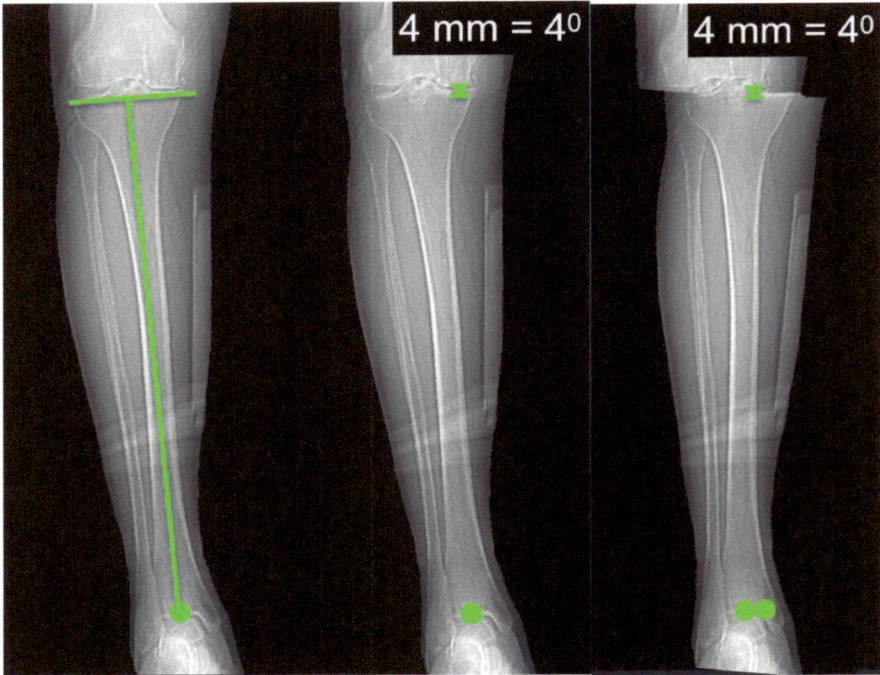

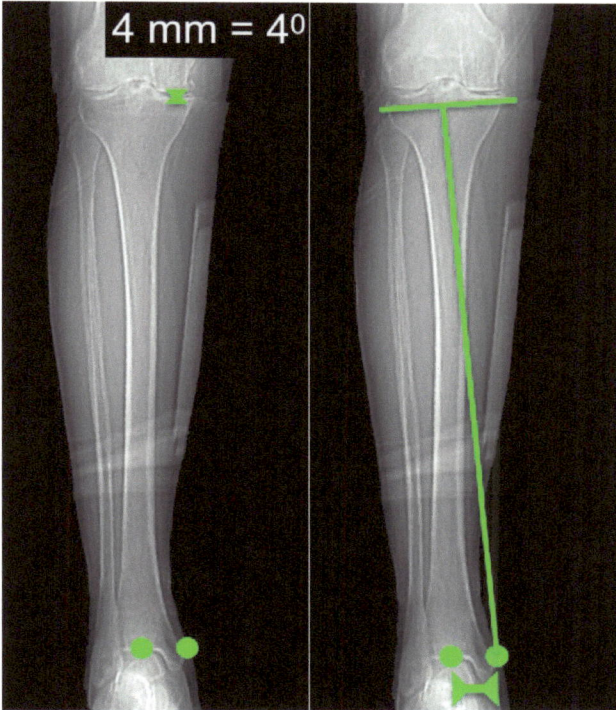

$4 \text{ mm} = 4^0$

This example of a varus knee deformity shows that compensating 2 mm for distal medial femoral cartilage loss and 2 mm for proximal medial tibial cartilage and bone loss will shim open the medial compartment 4 mm and restores the patient's pre-arthritic alignment. Four millimeters of correction at the joint line moves the ankle 24 millimeters lateral, correcting the deformity caused by varus osteoarthritis.

Consider following the 3rd Commandment:

3.   THOU SHALT NOT MAKE A STRAIGHT LEG INSTEAD STRAIGHTEN THE LEG TO THE PATIENT'S PRE-ARTHRITIC ALIGNMENT.

Chapter 4 The 4th Commandment

The 4th Commandment: **Thou Shalt Not Worry About Varus Failure of the Tibial Component as This is Caused by Mechanical Alignment and Alleviated by Kinematic Alignment**, is very controversial.

> ## 4.   THOU SHALT NOT WORRY ABOUT VARUS FAILURE OF THE TIBIAL COMPONENT AS THIS IS CAUSED BY MECHANICAL ALIGNMENT AND ALLEVIATED BY KINEMATIC ALIGNMENT.

Why is the incidence of tibial component failure after mechanically aligned TKA higher than kinematically aligned TKA?

> ## Why Does MA TKA Have High Rate of Tibial Component Failure?

One reason is there is a wide range of phenotypes of the native distal femur in the young non-osteoarthritic knee and mechanical alignment sets the femoral component in varus in most of the knees.

Phenotyping the knee in young non-osteoarthritic knees shows a wide distribution of femoral and tibial coronal alignment

Michael T. Hirschmann[1,3] · Lukas B. Moser[1,3] · Felix Amsler[5] · Henrik Behrend[4] · Vincent Leclercq[6] · Silvan Hess[1,2]

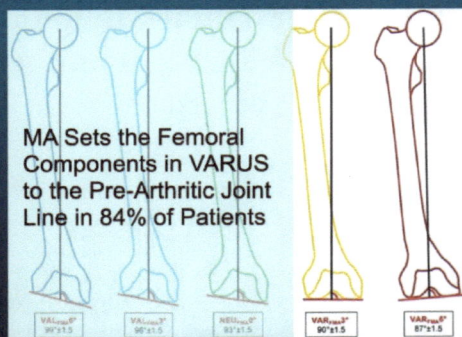

- MA sets nearly all femoral components in varus

- Calipered KA restores patient's pre-arthritic joint line and sets no femoral components in varus

MA Sets the Femoral Components in VARUS to the Pre-Arthritic Joint Line in 84% of Patients

Hirschmann, KSSTA, 2019

Mechanical alignment, striving for a joint line perpendicular to the femur's mechanical axis, sets 84% of the distal femur joint lines in varus relative to the patient's pre-arthritic increases the risk of varus overload at the knee. In contrast, calipered kinematic alignment restores the patient's pre-arthritic distal femoral joint line minimizing the risk of varus overload.

When mechanical alignment sets the femoral component in varus in 84% of patients requiring total knee replacement in varus, then the tibial joint line's orientation must be in an equal amount of valgus to compensate. There is a double varus deformity when there is a valgus under-compensation of the tibial component's orientation. Varus setting of the femoral component explains why mechanical and not kinematic alignment causes varus tibial component failure.

Another reason mechanical alignment has a higher risk of varus tibial component failure is that the knee adduction moment is higher than kinematic alignment.

Kinematically aligned total knee arthroplasty reduces knee adduction moment more than mechanically aligned total knee arthroplasty

Yasuo Niki[1] · Takeo Nagura[2] · Katsuya Nagai[1] · Shu Kobayashi[1] · Kengo Harato[1]

Received: 4 May 2017 / Accepted: 27 October 2017
© European Society of Sports Traumatology, Knee Surgery, Arthroscopy (ESSKA) 2017

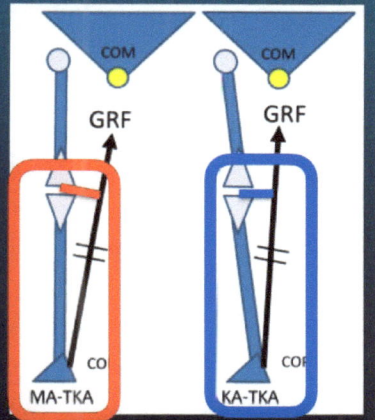

- KA is a promising option for patients with large coronal bowing of the tibia because the low knee adduction moment lowers the risk of varus tibial loosening

Niki, KSSTA, 2017

Mechanical alignment's change of the patient's pre-arthritic joint line widens the patient's stance during gait, which increases the adduction moment at the knee. The narrower stance after calipered kinematic alignment from restoring the patient's pre-arthritic gait lowers the knee's adduction moment and varus overload risk.

As these studies predicted, the risk of varus tibial overload after mechanical alignment TKA is high after implantation at six months, 1-year, and 2-years as measured by radiostereometric analysis (RSA).

The Effect of Coronal Alignment on Tibial Component Migration Following Total Knee Arthroplasty

A Cohort Study with Long-Term Radiostereometric Analysis Results

- A case-series of cherry-picked patients showed that MA TKA has higher risk of tibial component migration when alignment is mechanically out-of-range especially varus

van Hamersveld JBJS, 2019

After kinematically aligned TKA, failure from varus tibial overload is rare at two-years, according to a radiostereometric analysis Level 1 randomized trial.

■ KNEE

A randomized controlled trial of tibial component migration with kinematic alignment using patient-specific instrumentation versus mechanical alignment using computer-assisted surgery in total knee arthroplasty

E. K. Laende,
C. G. Richardson,
M. J. Dunbar

- A case-series of cherry-picked patients showed that MA TKA has higher risk of tibial component migration when alignment is mechanically out-of-range especially varus

- A RCT, at two-year follow-up, showed that tibial component migration after KA TKA is the same as MA TKA, and migration is NOT associated with post-operative alignment

van Hamersveld JBJS, 2019   Dunbar, BJJ, 2019

The tibial component migration after kinematically aligned TKA was negligible, as measured by radiostereometric analysis.

The following is an example of one of our patients where mechanical alignment set the femoral component in varus to the patient's distal femoral joint line (left), and kinematic alignment restored the patient's pre-arthritic distal femoral joint line (right). Varus tibial component failure occurred in the TKA set with mechanical alignment and not with kinematic alignment.

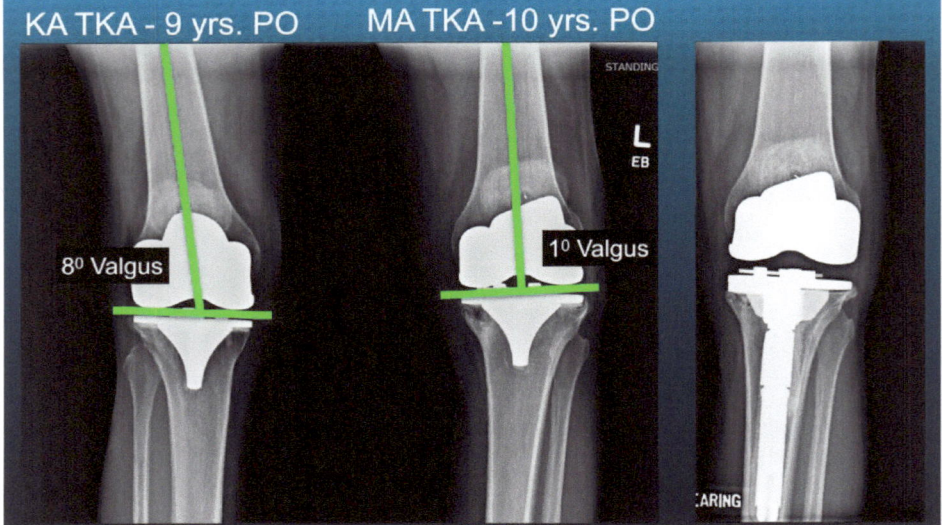

Femoral Component Set in Varus by MA, Caused Varus Tibial Component Failure

KA TKA - 9 yrs. PO   MA TKA -10 yrs. PO

8⁰ Valgus

1⁰ Valgus

The patient's right knee was kinematically aligned nine years ago, and the left knee was mechanically aligned ten years ago. On the right side, the distal lateral femoral angle is 8 degrees valgus and 1-degree valgus on the left side. The 7-degree more varus deviation of the left femoral component caused varus or medial overload. Hence, mechanical alignment causes varus failure, and kinematic alignment reduces the risk.

How does the superior biomechanics of a kinematically aligned TKA translate in implant survival and function ten years after kinematically aligned TKA?

Implant Survival and Function Ten Years After Kinematically Aligned Total Knee Arthroplasty

Stephen M. Howell, MD [a], Trevor J. Shelton, MD, MS [b, *], Maury L. Hull, PhD [a, b, c]

- 98.5% aseptic survival of 220 KA TKA performed in 2007 (Howell, J Arthro, 2018)

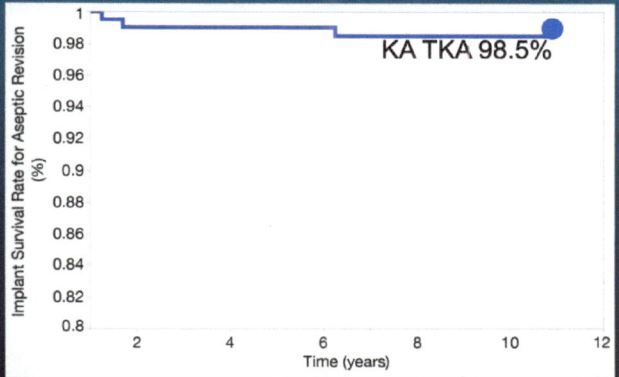

The implant survival rate for aseptic revision after calipered kinematic alignment is 98% at 10-years, as reported by a single surgeon in a consecutive series of 220 TKAs.

This survival is comparable if not higher than the implant survival of similar single-surgeon case-series of patients after mechanical alignment.

■ KNEE

The effect of post-operative mechanical axis alignment on the survival of primary total knee replacements after a follow-up of 15 years

- 98.5% aseptic survival of 220 KA TKA performed in 2007
(Howell, J Arthro, 2018)

- 93.7% aseptic survival of 270 MA TKA
(Bonner, BJJ, 2011)

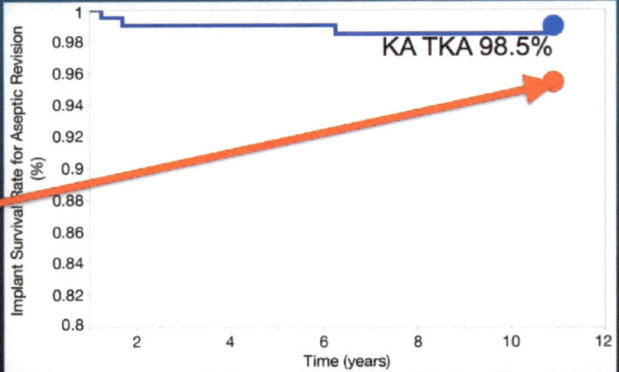

After mechanical alignment, the implant survival rate is 94% at 10-years, as reported by a single surgeon in a series of 270 TKAs performed in the United Kingdom.

Effect of Postoperative Mechanical Axis Alignment on the Fifteen-Year Survival of Modern, Cemented Total Knee Replacements

By Sebastien Parratte, MD, PhD, Mark W. Pagnano, MD, Robert T. Trousdale, MD, and Daniel J. Berry, MD

- 98.5% aseptic survival of 220 KA TKA performed in 2007
(Howell, J Arthro, 2018)

- 93.7% aseptic survival of 270 MA TKA
(Bonner, BJJ, 2011)

- ~93% aseptic survival of 398 MA TKA
(Parratte, JBJS, 2010)

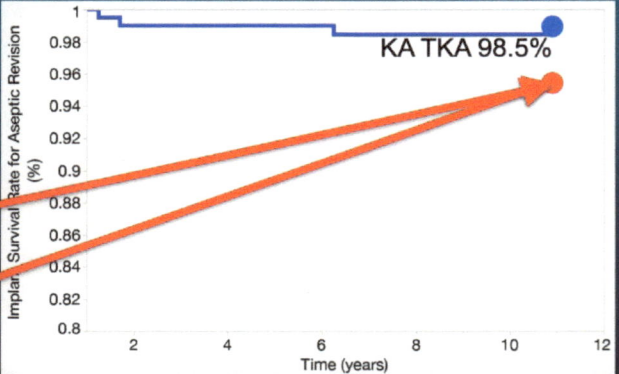

The implant survival rate is 93% at 10-years, as reported by a single surgeon of 398 mechanically aligned TKAs performed at the Mayo Clinic in the United States.

The difference in the implant survival rate between kinematically and mechanically aligned TKA results in significant health care savings when considering the cost of the number of revision surgeries.

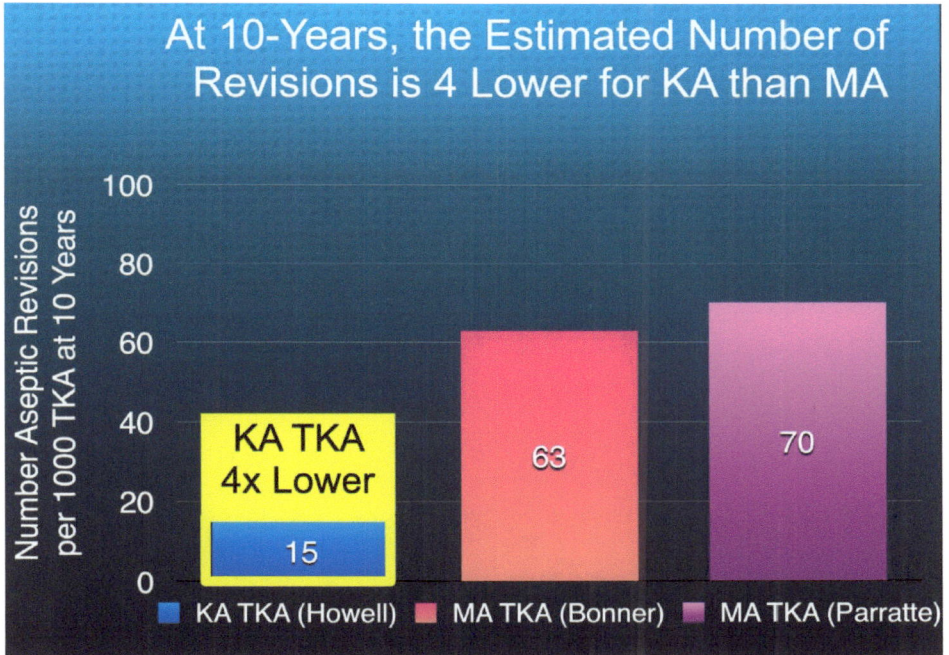

The projected rate of 15 revisions/1000 TKAs after calipered kinematically aligned TKA is four times lower than mechanical alignment with a rate of 63 and 70 revisions/1000 TKAs.

Consider following the 4th Commandment:

4. THOU SHALT NOT WORRY ABOUT VARUS FAILURE OF THE TIBIAL COMPONENT AS THIS IS CAUSED BY MECHANICAL ALIGNMENT AND ALLEVIATED BY KINEMATIC ALIGNMENT.

.

Chapter 5 The 5th Commandment

The 5th Commandment is **Thou Shalt Not Profess Allegiance to the Center of the Femoral Head and Ankle when Aligning a Total Knee Arthroplasty**.

5.   THOU SHALT NOT PROFESS ALLEGIANCE TO THE CENTER OF THE FEMORAL HEAD AND ANKLE.

Disregarding mechanical alignment targets can be very difficult for initiates beginning the transition from mechanical to calipered kinematic alignment. Following mechanical alignment targets are hazardous since the three kinematic axes of the knee have no relationship with the hip and ankle center locations.

Three Kinematic Axes Have No Alignment Relationship with the Hip and Ankle Centers

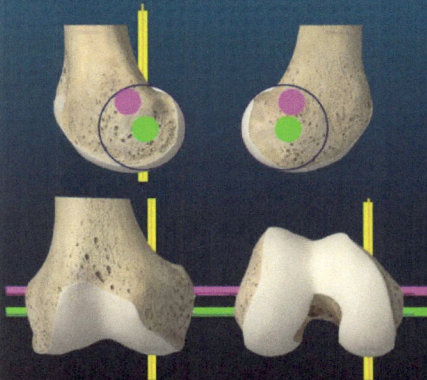

Medial Ball and Socket

This schematic shows a green line that is the axis about which the tibia flexes and extends, a magenta line that is the one about which the patella flexes and extends on the femur, and a yellow vertical line about which the tibia internally and externally rotates. These axes are either perpendicular or parallel to the native joint line.

Let's look at the adverse consequences of mechanical alignment, as Niki did. There is a very high risk of bony gap changes and displacement of the flexion-extension axes in the femur about which the tibia moves and about which the patella moves.

Mechanically aligned total knee arthroplasty carries a risk of bony gap changes and flexion–extension axis displacement

Yasuo Niki[1] · Tomoki Sassa[1] · Katsuya Nagai[1] · Kengo Harato[1] · Shu Kobayashi[1] · Taro Yamashita[1]

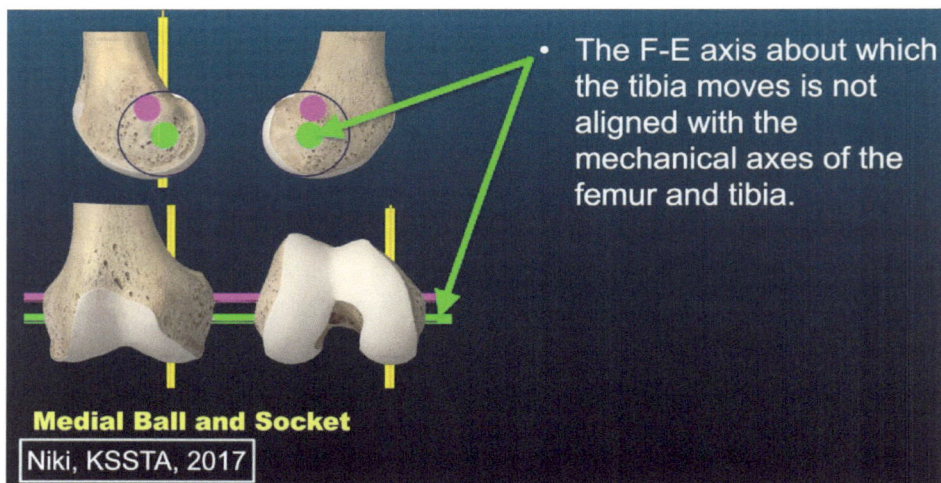

• The F-E axis about which the tibia moves is not aligned with the mechanical axes of the femur and tibia.

Medial Ball and Socket

Niki, KSSTA, 2017

As there is no relationship between the orientation of the flexion-extension axes about which your tibia and patella moves and the mechanical axes of the femur and tibia, coalignment of the axes of the components with those of the knee cannot occur using the center of the hip and the center of the ankle.

Similarly, Iranpour looked at the trochlear groove's geometry and the relationship to the trans-epicondylar axis, a commonly used reference for mechanical alignment.

The Geometry of the Trochlear Groove

**Farhad Iranpour MD, Azhar M. Merican BM, MS(Orth),
Wael Dandachli MRCS, Andrew A. Amis DSc(Eng),
Justin P. Cobb FRCS**

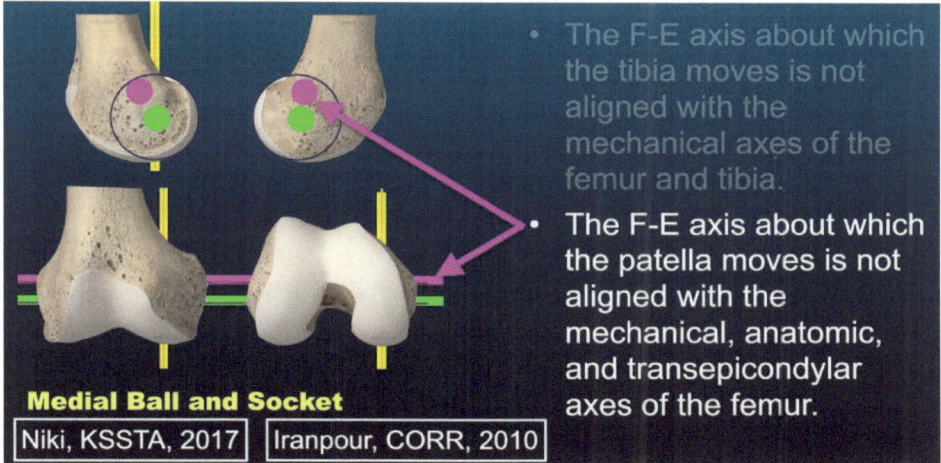

- The F-E axis about which the tibia moves is not aligned with the mechanical axes of the femur and tibia.

- The F-E axis about which the patella moves is not aligned with the mechanical, anatomic, and transepicondylar axes of the femur.

Medial Ball and Socket

Niki, KSSTA, 2017   Iranpour, CORR, 2010

They showed that the flexion-extension axis about which the patella moves deviates from the femur's transepicondylar axis.

By 2019 several studies showed that the femoral head and ankle targets in the UKA and TKA are not the gold alignment standard.

2019 Studies Show the Target of the Femoral Head and Ankle in UKA and TKA is Not the Gold Standard

The study of revision rate and functional outcome after Oxford medial unicompartmental knee replacement showed no relationship to the final limb alignment.

Functional Outcome and Revision Rate Are Independent of Limb Alignment Following Oxford Medial Unicompartmental Knee Replacement

J.A. Kennedy, MBBS, MRCS, J. Molloy, MD, C. Jenkins, MPhil, S.J. Mellon, BSc(Hons), PhD, C.A.F. Dodd, FRCS(Orth), and D.W. Murray, MD, FRCS(Orth)

- "These data support the standard operative technique for the Oxford UKR, which aims to restore ligament tension and therefore pre-arthritic alignment (i.e. KA) rather than neutral mechanical alignment"

Kennedy, JBJS, 2019

The data supported the Oxford UKR standard operating technique to restore ligament tension and, therefore, pre-arthritic joint lines like kinematic alignment rather than mechanical alignment.

There was no better implant survival and functional outcome when computer-assisted surgery more accurately aligned the knee to mechanical alignment targets, according to a review of 19,221 total knee arthroplasties.

Outcomes of Computer-Assisted Surgery Compared with Conventional Instrumentation in 19,221 Total Knee Arthroplasties

Results After a Mean of 4.5 Years of Follow-Up

Timothy D. Roberts, MBChB, Christopher M. Frampton, BSc(Hons), PhD, and Simon W. Young, MBChB, FRACS, MD

- "These data support the standard operative technique for the Oxford UKR, which aims to restore ligament tension and therefore pre-arthritic alignment (i.e. KA) rather than neutral mechanical alignment"
- "The present study demonstrated no difference in survivorship or functional outcome scores to support using CAS for TKA"

Kennedy, JBJS, 2019   Roberts, JBJS, 2019

The paradigm shifted to calipered kinematic alignment since there is no benefit from 'better' alignment of the total knee arthroplasty to the femoral head and ankle center.

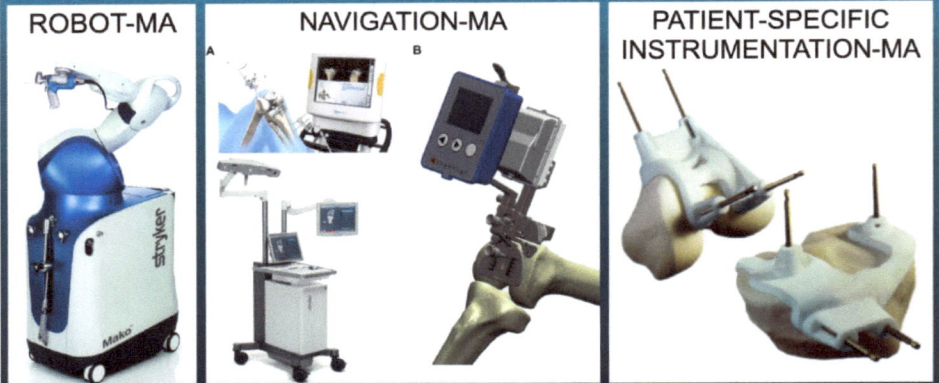

The Alignment Paradigm Has Shifted to Calipered KA

ROBOT-MA	NAVIGATION-MA	PATIENT-SPECIFIC INSTRUMENTATION-MA

- Using these technologies to perform MA adds expense and time and no benefit in terms of implant survival, patient satisfaction, and clinical outcomes

The consensus when using a robot, navigation, and patient-specific instrumentation to perform mechanical alignment is that these techniques add expense, complications, and operative time with no proven benefit in terms of implant survival patient satisfaction and clinical outcomes.

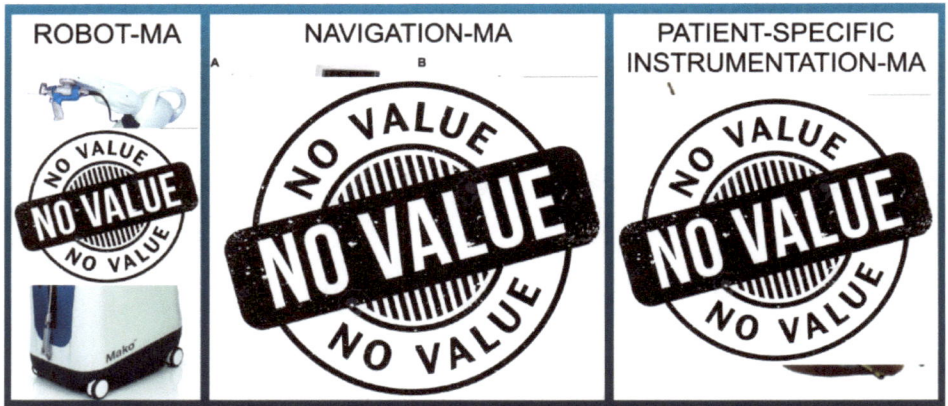

ROBOT-MA	NAVIGATION-MA	PATIENT-SPECIFIC INSTRUMENTATION-MA

Those who expect value from using robot, navigation, and patient-specific instrumentation should use instruments to perform calipered kinematic alignment and not mechanical alignment. We stress the need to double-check that the bone's resections match the intended kinematic alignment target with a caliper.

So if you're cutting with a robot, navigation, and patient specific instrumentation take a $5 caliper and measure the thickness of the distal and posterior resections of your femur and make sure that they match those of the prosthetic femoral condyle after you compensate for a millimeter of the curve of the blade and two millimeters when there is cartilage missing.

Consider following the 5th Commandment:

5.   THOU SHALT NOT PROFESS ALLEGIANCE TO THE CENTER OF THE FEMORAL HEAD AND ANKLE.

Chapter 6 The 6th Commandment

The 6th Commandment is **Thou Shalt Not Deny Treatment to Knees with Severe Varus, Valgus, and Flexion Preoperative Deformities.**

6.   THOU SHALT NOT DENY TREATMENT TO KNEES WITH SEVERE VARUS, VALGUS, AND FLEXION DEFORMITIES.

Beginning the transition from mechanical to kinematic alignment can be daunting. Early on, there are some knees where the post-operative component orientation after kinematic alignment might make you uncomfortable. We suggest you use mechanical alignment in these few knees until you gain experience. With time, if your experience parallels ours after performing over 13,500 TKAs with calipered kinematic in all-comers since 2006, you will treat even the most severe deformities with kinematic alignment.

The following is a typical example of a challenging case.

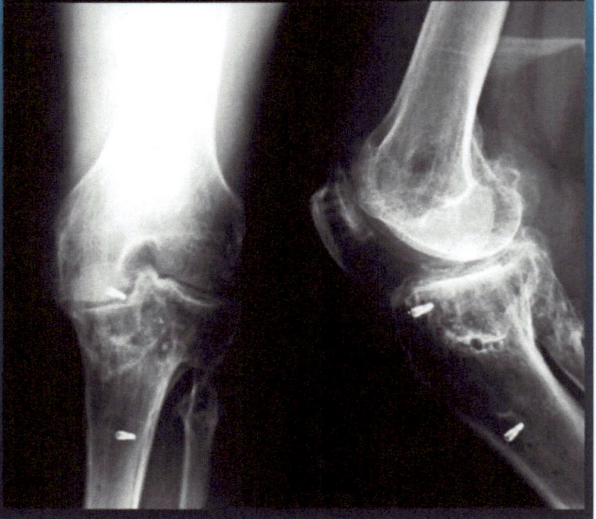

Medial Depressed Tibial Plateau Fracture and Osteoarthritis

- 57 y/o male
- Is the medial collateral ligament contracted?
- Is a medial augment needed?

The depression of this medial tibial plateau fracture might suggest the tibial bone loss would need an augment and stem extension.

There is generally no bone loss on the femur at 0 and 90 degrees of flexion where manual kinematic alignment manual instruments reference the femur.

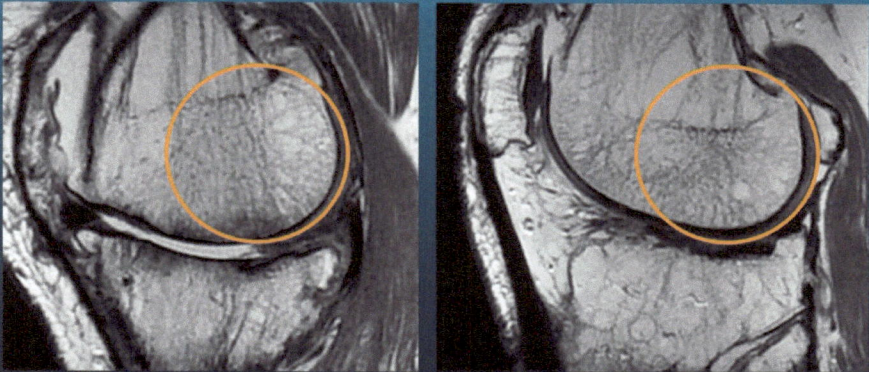

Is Cartilage or Bone Worn and How Much?

- Medial and lateral femoral condyles have same radii

Howell, JBJS, 2010; Nam, KSSTA, 2014

When you need convincing, get an MRI and fit the same size best fit circle from 15 to 120 degrees of flexion and lateral and medial femoral condyle, which will show no real bone loss. Here cartilage loss is confined to the distal medial femur, with no cartilage loss posterior medial, distal lateral, and posterior lateral.

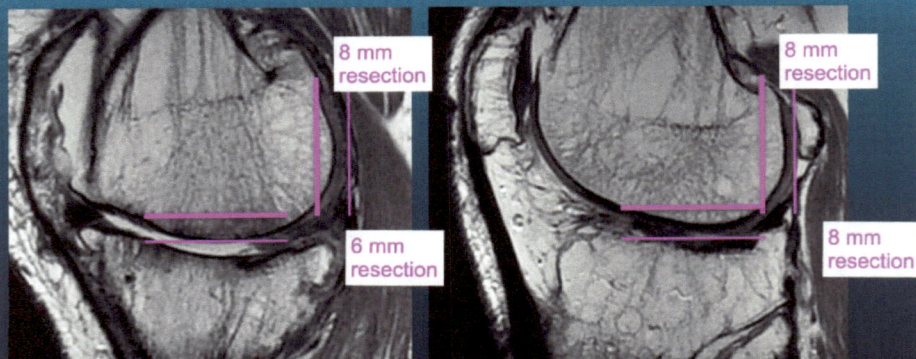

Is Cartilage or Bone Worn and How Much?

8 mm resection

8 mm resection

6 mm resection

8 mm resection

- Medial and lateral femoral condyles have same radii

Howell, JBJS, 2010; Nam, KSSTA, 2014

When you measure the resection with a caliper after accounting for the 1 mm kerf equal to the saw blade's thickness, the resections should be 6 mm distal medial and 8mm thick for the remaining three resections.

Was the Femoral Component Positioned Kinematically?

DM 6.5   PM 5+3   DL 8.5   PL 8

2 pieces

- Initial posterior medial bone resection was 5 instead of 8 mm

Sometimes the resections need fine-tuning, which the caliper measurements determine. In this case, the distal medial target was 6 mm, and the resection was 6.5 mm-close enough. The posterior medial resection was too thin at 5 mm, so 3 mm more was resected, which yielded two pieces that met the 8 mm target. The distal lateral resection was 8.5 mm the posterior lateral resection was 8 mm, both within the 0.5 mm target. So, the caliper is your tool! We corrected the deformity on the femoral side to the patient's pre-arthritic distal and posterior femoral joint.

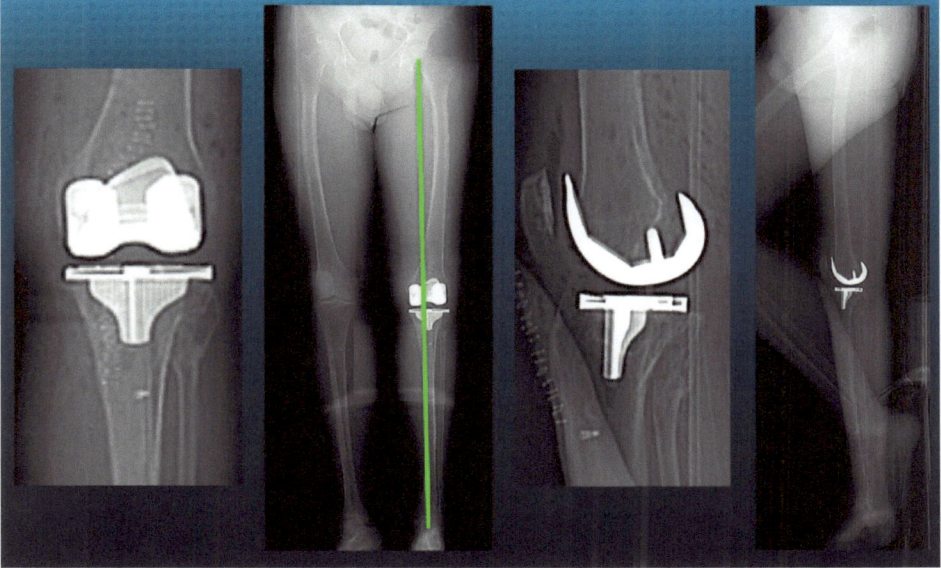

Calipered KA TKA Corrected The Severe Varus Deformity Without Ligament Release

These images show the post-operative alignment of the knee and limb alignment is acceptable. However, because the right leg's rotation does not match the left, the joint lines' ideal comparison of the contralateral pre-arthritic knee is impossible.

The following is a typical example of a knee treated with a valgus osteoarthritic deformity with calipered kinematic alignment TKA.

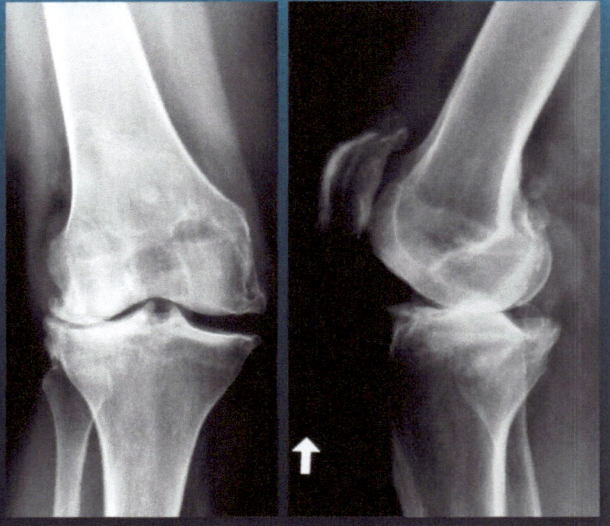

Calipered KA Corrects Severe Valgus Deformities Without Ligament Release

- 50 y/o female
- Is the medial collateral ligament stretched?
- Is the LFC hypoplastic?
- Implant concerns

The valgus deformity can perplex the surgeon transitioning to calipered kinematic alignment because of the dogma the MCL stretched, and there is a hypoplasia of the lateral femoral condyle. The medial compartment's exaggerated opening results from a chronic posterolateral capsular contracture that occurs from imaging the knee in flexion. The medial side opens a few more millimeters because the medial collateral ligament is slacker with the knee is flexion than in extension.

An MRI study convincingly showed no hypoplasia of the lateral femoral condyle in the knee with a valgus osteoarthritic deformity.

Assessment of the Radii of the Medial and Lateral Femoral Condyles in Varus and Valgus Knees with Osteoarthritis

By Stephen M. Howell, MD, Stacey J. Howell, and Maury L. Hull, PhD

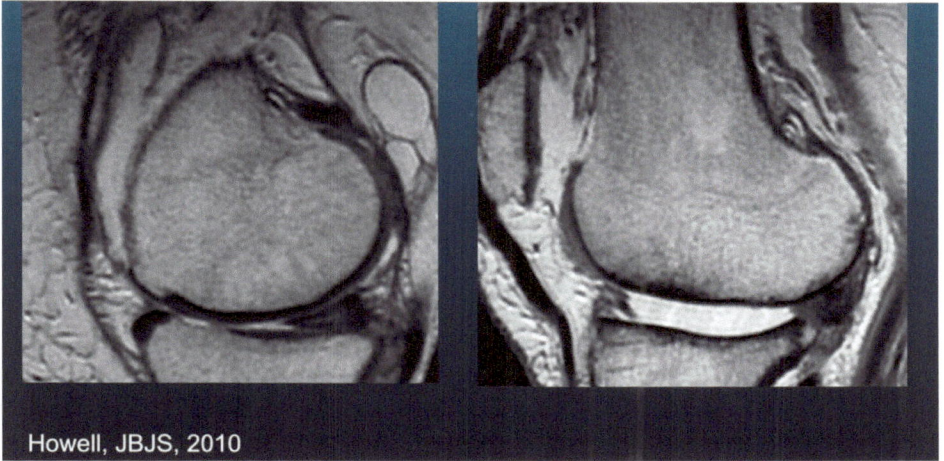

Howell, JBJS, 2010

The medial and lateral femoral condyles' radii were similar in 44 valgus knees with osteoarthritis treated with a primary total knee replacement.

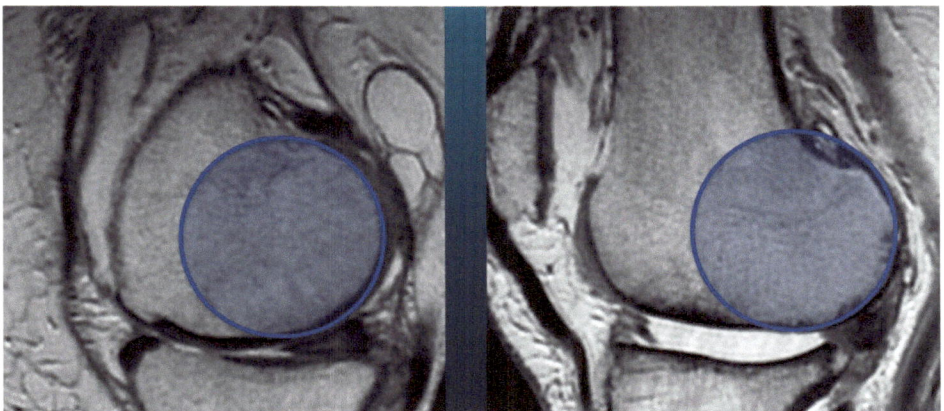

There was no difference between the radii of a circle fit between the lateral and medial femoral condyle. Verifying the lateral femoral condyle is not hypoplastic relative to the medial femoral condyle.

So, proximal translation of the lateral femoral condyle relative to the medial one creates the valgus knee. The radius is not smaller.

You can plan the thickness of the resections you will measure with a caliper for a femoral component with 9mm thick distal and posterior femoral condyles.

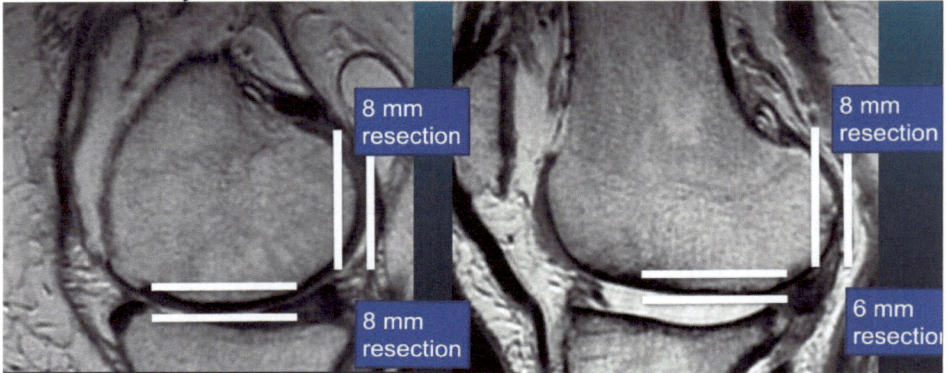

When you measure with a caliper after accounting for the 1 mm kerf of a saw blade's thickness, the resections should be 6 mm distal lateral and 8mm thick for the remaining three resections.

Bone Resections Confirmed Femoral Component Was KA

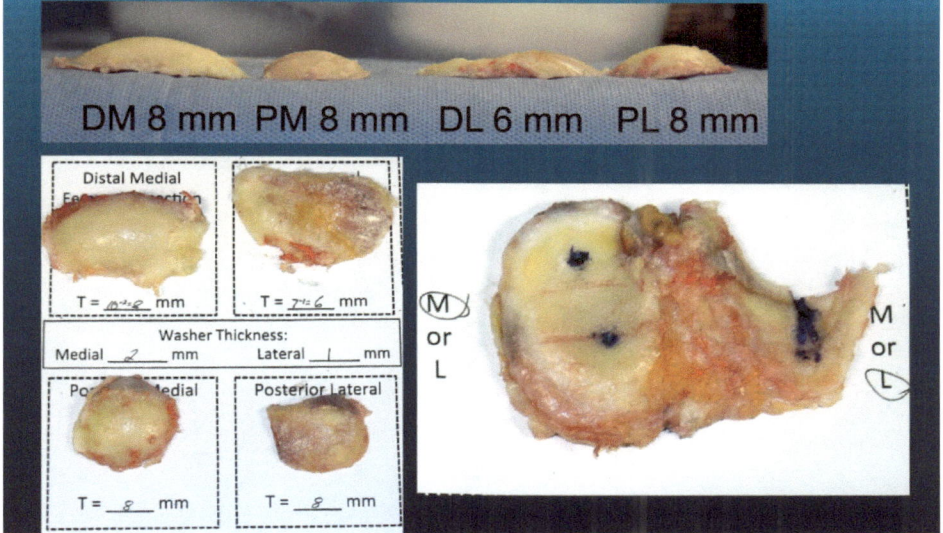

DM 8 mm PM 8 mm DL 6 mm PL 8 mm

Distal Medial
T = 8.2 mm

T = 7.6 mm

Washer Thickness:
Medial __2__ mm   Lateral __1__ mm

Posterior Medial
T = __8__ mm

Posterior Lateral
T = __8__ mm

(M) or L

M or L

The distal medial and lateral resections were too thick. Shimming with a 2 mm thick washer medially and 1 mm thick washer laterally placed on the four-in-one chamfer block's posts created gaps eventually filled with cement the corrected the over-resection.

There was complete erosion of cartilage and several millimeters of bone on the posterolateral tibia, resulting in severe valgus deformity. The tibial component setting compensated for these defects, corrected the deformity, and restored the patient's pre-arthritic alignment.

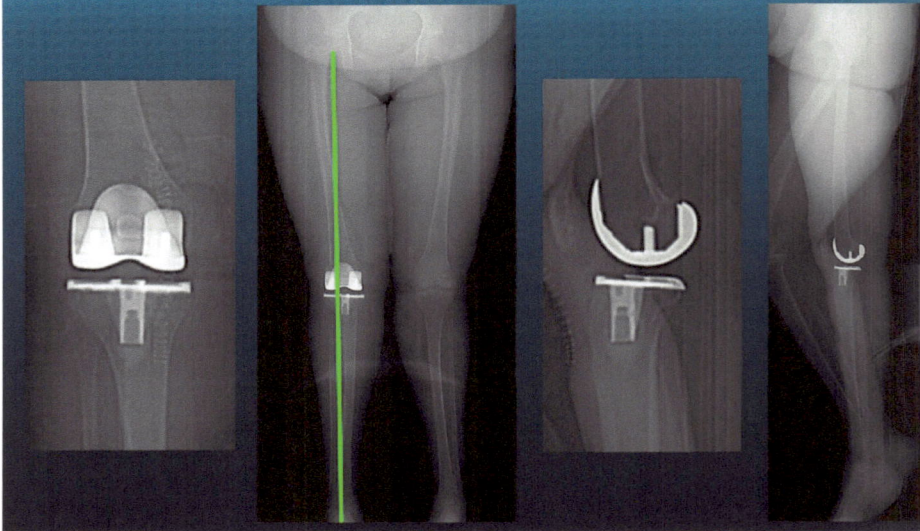

Limb and Joint Lines Restored to Patient's Pre-Arthritic and Contralateral Alignment

These images show the post-operative alignment of the knee and limb alignment is acceptable as the level and orientation of the distal femoral and proximal tibial joint lines are comparable to those of the contralateral pre-arthritic knee.

The calipered kinematic alignment technique uses the same principle of restoring the patient's pre-arthritic joint lines to treat severe varus and valgus deformities successfully.

Consider following the 6th Commandment:

6.   THOU SHALT NOT DENY TREATMENT TO KNEES WITH SEVERE VARUS, VALGUS, AND FLEXION DEFORMITIES.

Chapter 7 The 7th Commandment

The 7th Commandment is **Thou Shalt Not Commit to the First Tibial Cut Instead Use a Recut Guide to Optimize Varus-Valgus Stability in Extension and to Restore the Native Laxity in Flexion.**

> # 7. THOU SHALT NOT COMMIT TO THE FIRST TIBIAL CUT INSTEAD USE A RECUT GUIDE TO OPTIMIZE VARUS-VALGUS STABILITY IN EXTENSION AND RESTORE NATIVE LAXITY IN FLEXION.

We will show you the surgical steps and verification checks to achieve these goals. The first step is to place the knee in extension—insert a spacer block between the femoral and tibial resection, and then visually examine whether there's negligible varus-valgus laxity, which would indicate a snug rectangular space.

Place Knee in Extension, Insert Spacer Block, & Verify Negligible V-V Laxity Indicating a Rectangular Space

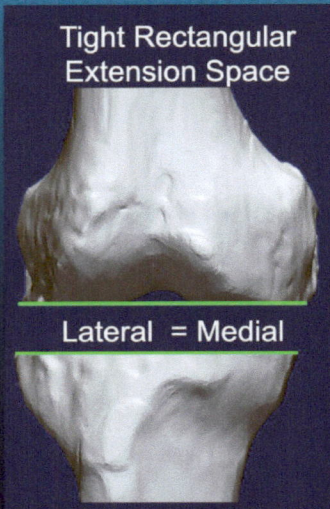

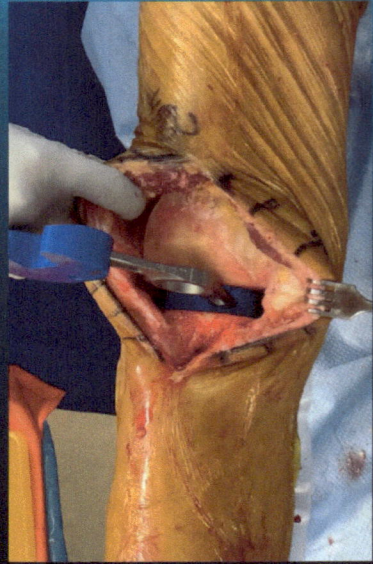

Tight Rectangular Extension Space

Lateral = Medial

Roth, JBJS, 2015; Roth JOR, 2018

We know from our published studies that all knees in full extension have negligible varus-valgus laxity, which explains why the knee is considered a rigid body in extension. The lateral and medial gaps should be the same, which would give a snug rectangular extension space.

When the medial side is too tight and the lateral side gaps, you need a varus recut to establish a rectangular space. When the lateral side is too tight and the medial side gaps, you need a valgus recut. How much of a recut? Look at how much bone separates from the spacer block, and when the compartment opens 1-2 millimeters, then a 1-degree recut is corrective.

When the compartment opens 3-4 millimeters, then a 2-degree recut is good. So, when the medial or lateral gap is too tight in extension, consult the decision tree.

When the Medial or Lateral Gap is Too Tight in Extension Consult Decision-Tree

DECISION-TREE FOR BALANCING A CALIPERED KINEMATICALLY ALIGNED MEDACTA **GMK SPHERE CR** TKA

Tight in Flexion & Extension	Tight in Flexion Well-Balanced in Extension	Tight in Extension Well-Balanced in Flexion	Well-Balanced in Extension and Loose in Flexion	Tight Medial & Loose Lateral in Extension	Tight Lateral and Loose Medial in Extension
Recut tibia and remove 1-2mm more bone.	Increase posterior slope until exposure A-P offset is restored at 90° of flexion.	Remove posterior osteophytes. Strip posterior capsule. Insert trial components & gently manipulate knee into extension.	Add thicker insert and recheck knee extends fully. When knee does not fully extend check PCL tension. When PCL is incompetent use **GMK Sphere CS** Insert.	Remove medial osteophytes. Reassess. Recut tibia in 1-2° more varus. Insert 1 mm thicker insert.	Remove lateral osteophytes. Reassess. Recut tibia in 1-2° more valgus. Insert 1 mm thicker insert.

There is an economy of options with calipered kinematic alignment because treatment decisions for balancing the knee are all based on adjusting the varus-valgus, slope, and thickness of the tibial resection.

Tight Medial & Loose Lateral in Extension

Remove medial osteophytes.

Reassess.

Recut tibia in 1-2° more varus.

Insert 1 mm thicker insert.

Tight Medial & Loose Lateral in Extension

Remove medial osteophytes.

Reassess.

Recut tibia in 1-2° more varus.

Insert 1 mm thicker insert.

Techniques for Fine-Tuning V-V Plane of Tibial Resection in 1⁰ to 2⁰ Increments

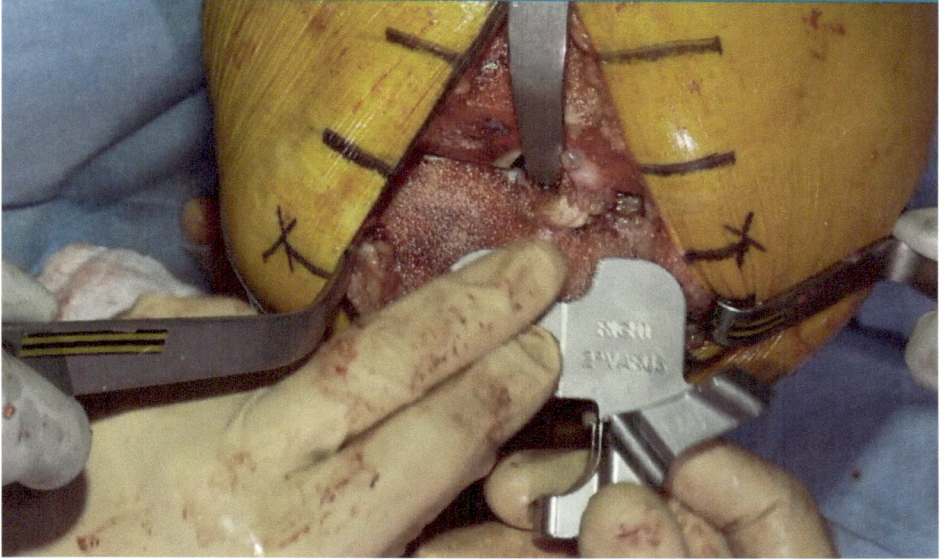

This schematic shows the resection of the right proximal tibia and the placement of a 2-degree varus recut guide to correct an overtight medial compartment, removing ~ 3 millimeters of bone off the medial tibial plateau.

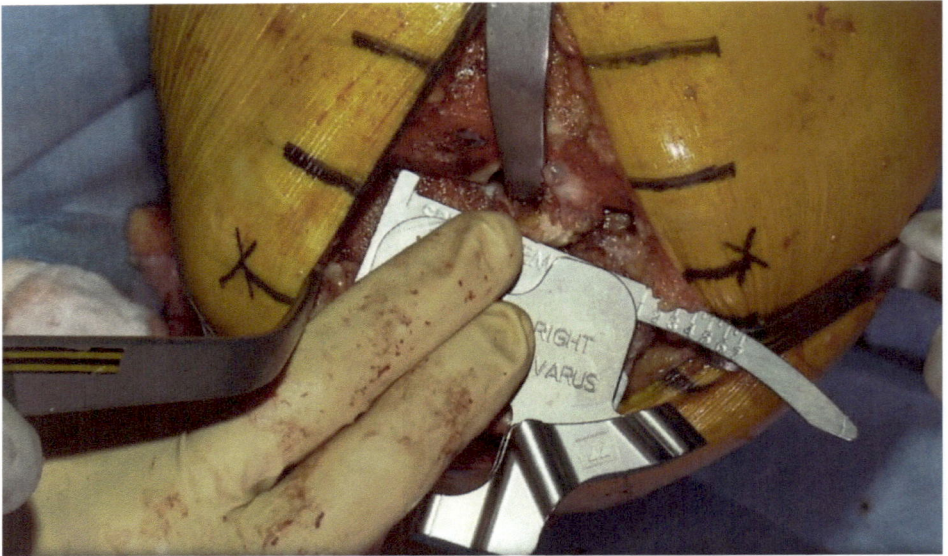

When less medial tibia needs resection, a 1 mm thicker angel wing on the proximal tibial resection and then 2-degree varus recut on top of it removes ~ 1-2 millimeters of bone off the medial tibial plateau.

Tight in Flexion
Well-Balanced in
Extension

Increase posterior
slope until exposure
A-P offset is
restored at 90°

When the TKA is tight in flexion and well balanced in extension, the decision-tree states the only explanation is too little posterior slope. Adding increments of posterior slope loosens the flexion space without changing the extension space.

As a rule, for every degree increase in posterior slope, the tibia moves posteriorly about 1 and ½ millimeters. This translation is measured with the offset caliper with the knee in 90 degrees of flexion as a change in the anterior-posterior offset of the tibia relative to the medial distal femur.

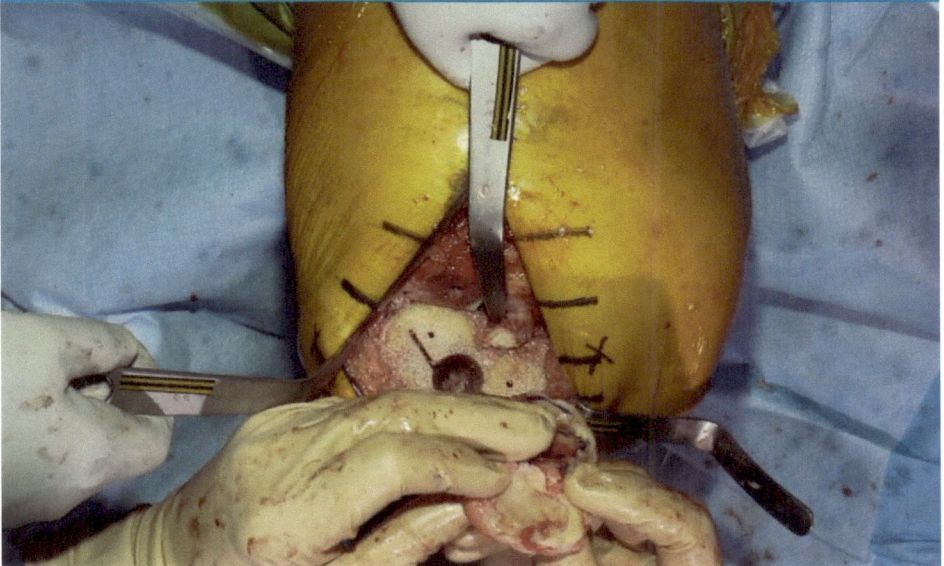

Fine-Tune Posterior Tibial Slope in 1-2⁰ Increments Until Native Slope is Restored

Examine the medial side of the tibial resection and determine whether the patient's pre-arthritic slope was restored. In this case, the slope was a degree or two too little, and a recut increased the posterior slope by shaving 1-2 mm of bone from the posterior tibia with the saw.

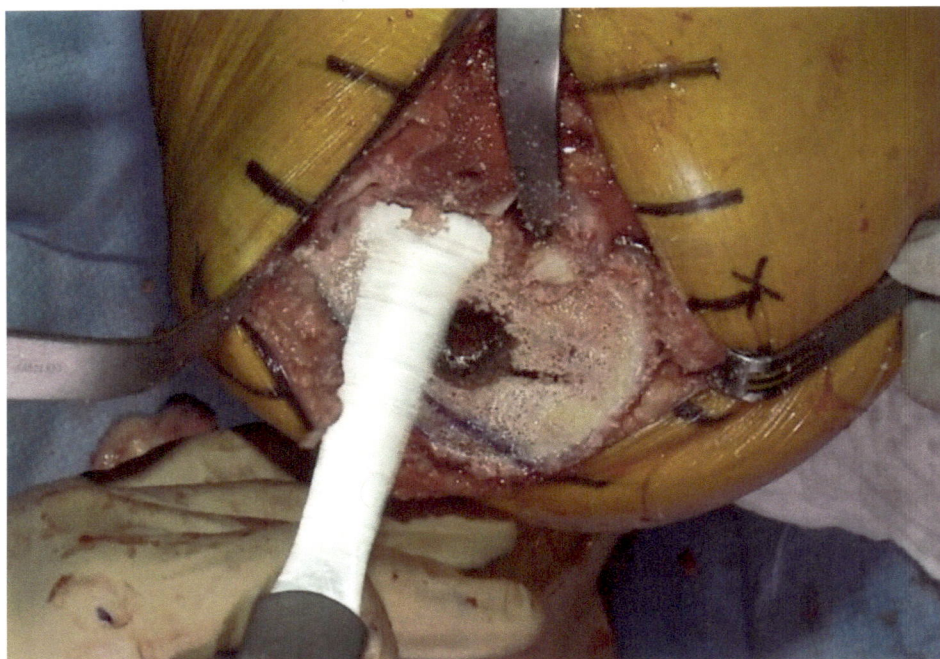

This schematic shows the saw shaving the posterior tibia to increase the slope using the blade's 1.2 mm thickness to gauge the resection thickness.

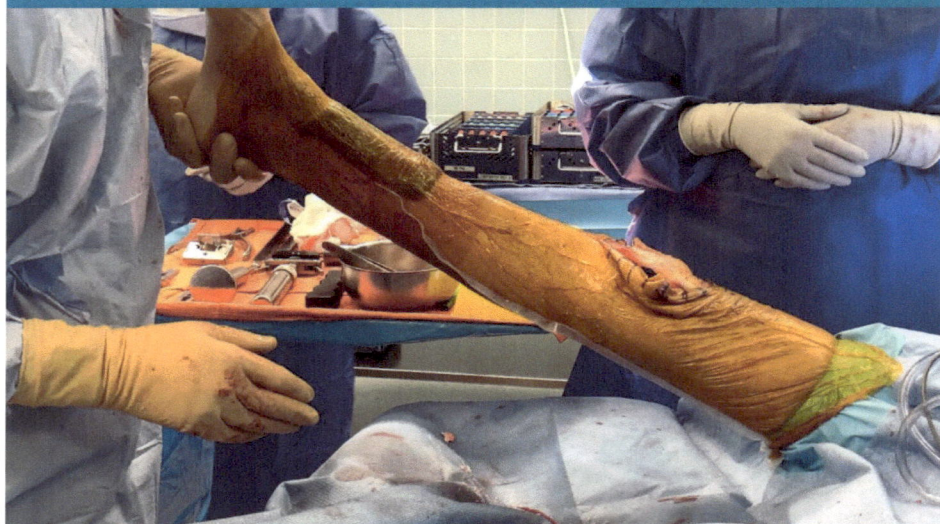

With Trial Components, Verify TKA Hyperextends Like Native Knee

Make the final assessment of balance with trial components. Be sure
the knee freely extends at least to zero degrees, or preferably a few
degrees of hyperextension, without recoiling into flexion.
s

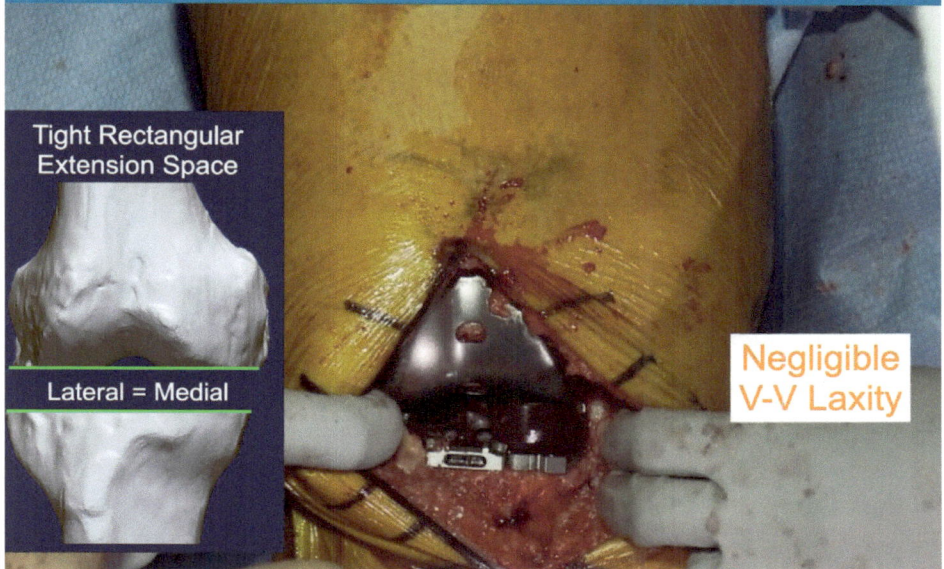

With Trial Components, Verify Negligible V-V Laxity in Full
Extension Which Restores Native Laxity of Extension Space

Tight Rectangular
Extension Space

Lateral = Medial

Negligible
V-V Laxity

With the knee in maximum extension, check that there is negligible
V.V. laxity. This medial ball-in-socket implant design (GMK Sphere,
Medacta) is stable in the sagittal and coronal planes.

Check the V.V. laxity with the knee in 15 to 30 degrees of flexion.
Based on laxity measurements in normal cadaveric knees, the lateral
gap is looser than the medial gap. Hence, expect the lateral gap opens
~ 3 mm, and the medial gap should be snug.

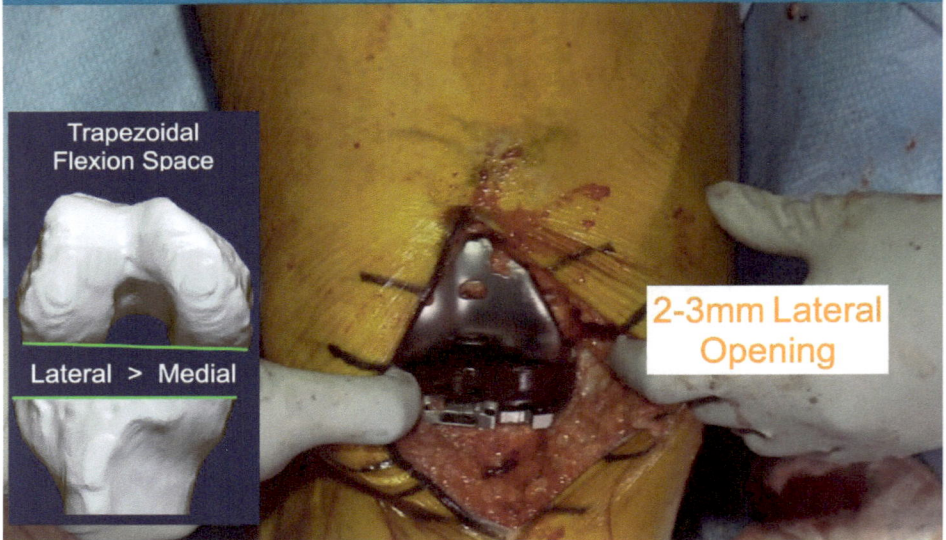

With Trial Components, Verify 2-3 mm of Lateral Opening with TKA in 15-30° of Flexion

Trapezoidal Flexion Space

Lateral > Medial

2-3mm Lateral Opening

Surgeons that perform arthroscopy see the asymmetric lateral and medial gaps all the time, which explains why removal of a torn lateral meniscus is easier than a torn medial meniscus.

The concept of gap balancing that strives to create a rectangular extension and flexion space is un-physiologic. With calipered kinematic alignment, the flexion space should be trapezoidal like the native knee.

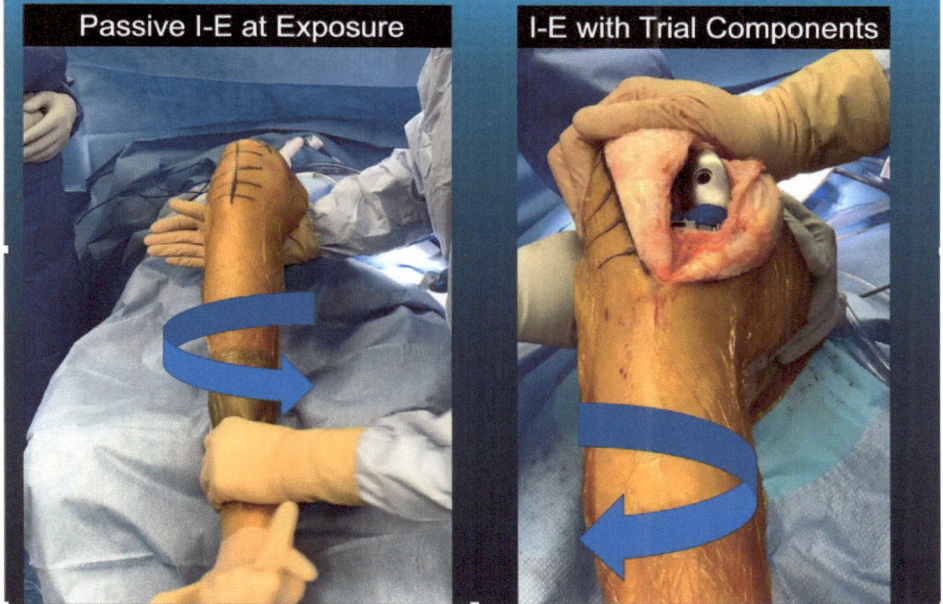

With Trial Components, Verify ± 15⁰ of I-E Rotation with TKA in 90⁰ of Flexion

Passive I-E at Exposure

I-E with Trial Components

Balancing the calipered kinematically aligned TKA requires the restoration of the native internal-external passive rotation and anterior-posterior offset with the knee in 90 degrees of flexion. Assessing the rotation before the surgery provides a baseline for the rotation to get with trial components. If it doesn't rotate, it's too tight. If it rotates too much, it's too loose.

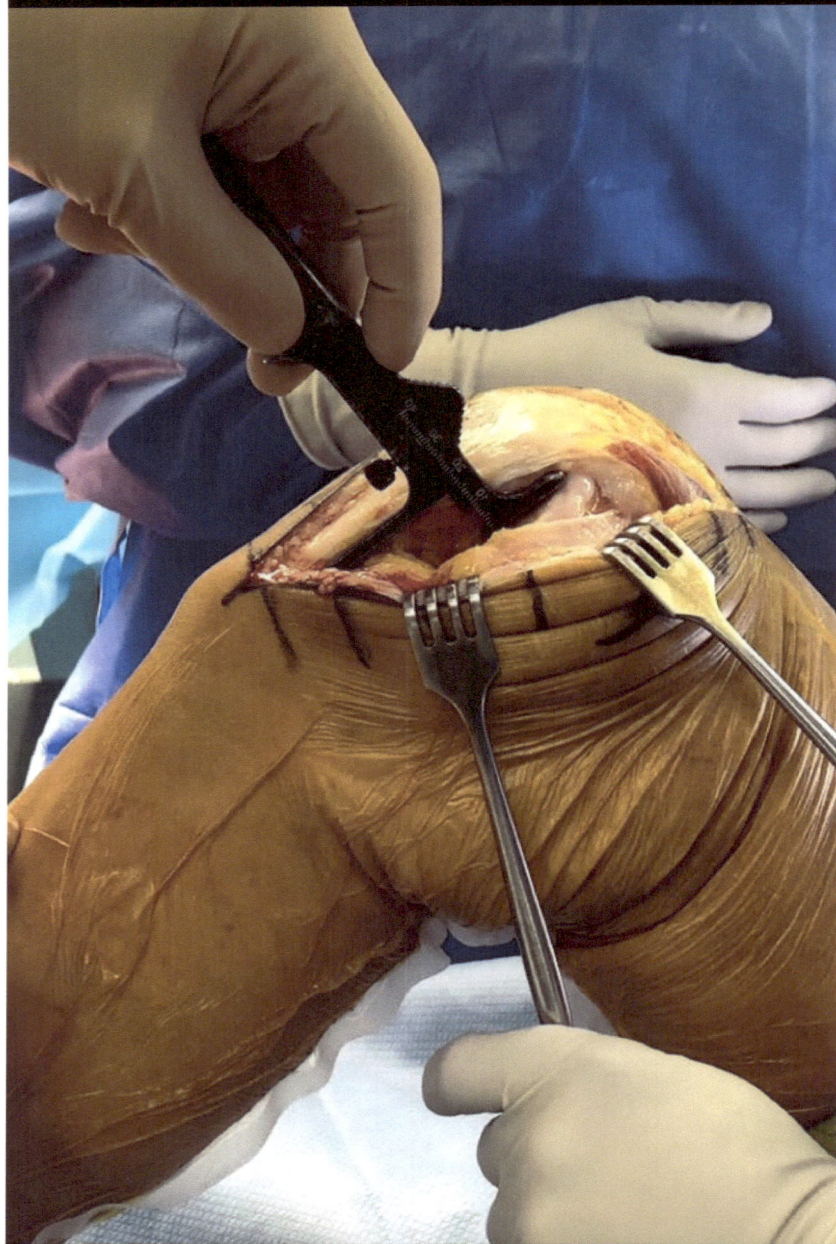

AP Offset at Exposure

16mm - 2mm = 14 mm

At the time of exposure, the offset of the anterior tibia from the distal femur in this varus knee was 16 mm, with the knee in 90 degrees of flexion when measured with the offset caliper.

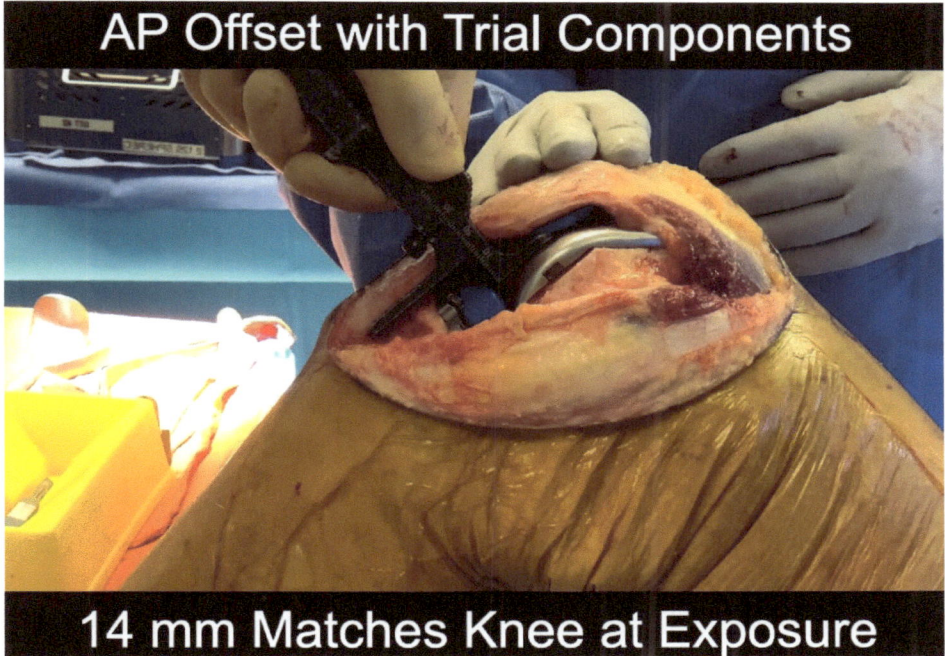

AP Offset with Trial Components

14 mm Matches Knee at Exposure

After compensating for 2 mm of cartilage wear on the medial side, the offset should be 14 mm with trial components when using a posterior cruciate ligament retaining insert. When the offset is greater than 14 mm, add more posterior slope. When the offset is less than 14 mm, then confirm the posterior cruciate ligament is intact and use a thicker insert or reduce the posterior slope.

Consider following the 7th Commandment:

7. THOU SHALT NOT COMMIT TO THE FIRST TIBIAL CUT INSTEAD USE A RECUT GUIDE TO OPTIMIZE VARUS-VALGUS STABILITY IN EXTENSION AND RESTORE NATIVE LAXITY IN FLEXION.

Chapter 8 The 8th Commandment

The 8th Commandment is **Thou Shalt Not Flex the Femoral Component More Than a Few Degrees from the Sagittal Anatomic Axis to the Distal Femur or Risk Patellofemoral Instability**.

8. THOU SHALT NOT FLEX THE FEMORAL COMPONENT MORE THAN A FEW DEGREES OR RISK PATELLOFEMORAL INSTABILITY.

A retrospective review of clinical characteristics and radiographic parameters showed that the femoral component's excessive flexion caused patellofemoral instability after a kinematically aligned total knee arthroplasty.

What clinical characteristics and radiographic parameters are associated with patellofemoral instability after kinematically aligned total knee arthroplasty?

Alexander J. Nedopil[1] · Stephen M. Howell[2] · Maury L. Hull[3]

Nedopil, Int Orthop, 2017

This review of 3,212 consecutive calipered kinematically aligned total knee replacements showed that 13 or 0.4% presented with patellofemoral instability.

In nine, the inciting event was an atraumatic subluxation that occurred at a mean of five months post-operatively. For example, one patient said while walking the dog, a pull on the leash rotated their leg and initiated the first patella subluxation. Another said they turned while making the bed, and the patella shifted.

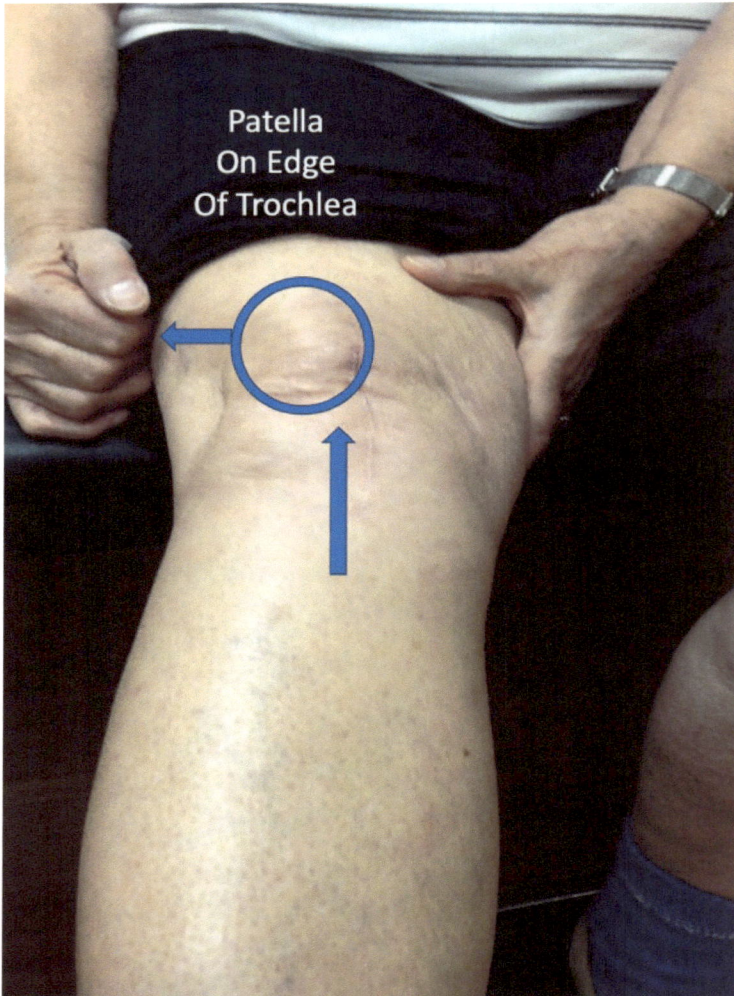

Patella
On Edge
Of Trochlea

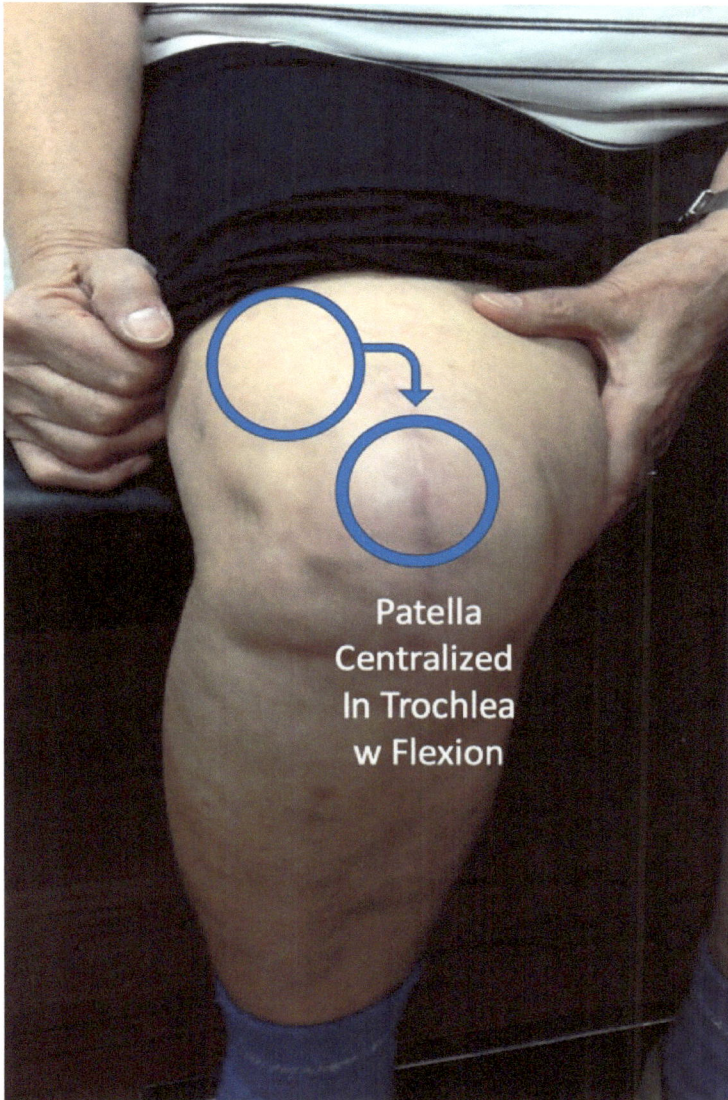

Patella
Centralized
In Trochlea
w Flexion

The photograph shows an active knee extension causing the quadriceps to pull the patella lateral in full extension lateral to the trochlea's edge (left). When the knee flexes, the patella shifts medial and centers in the trochlea. This instability is curious to the surgeon as it only happens with active motion and not passive motion. When you examine this patient and passively move the knee, it doesn't occur, making it difficult to confirm whether a revision stabilizing surgery was successful when the patient is still under anesthesia.

Patients with PF Instability Had 5° More Flexion of the Femoral Component Than Controls

A case-match study showed the cause of the patellofemoral instability by matching each patient with instability to three others randomly selected based on sex, age, preoperative knee deformity (i.e., varus or valgus), and the same implant design. The only difference between patients with patellofemoral instability and controls was that those with instability had a mean of six degrees more flexion of the femoral component relative to the anatomic axis of the distal femur.

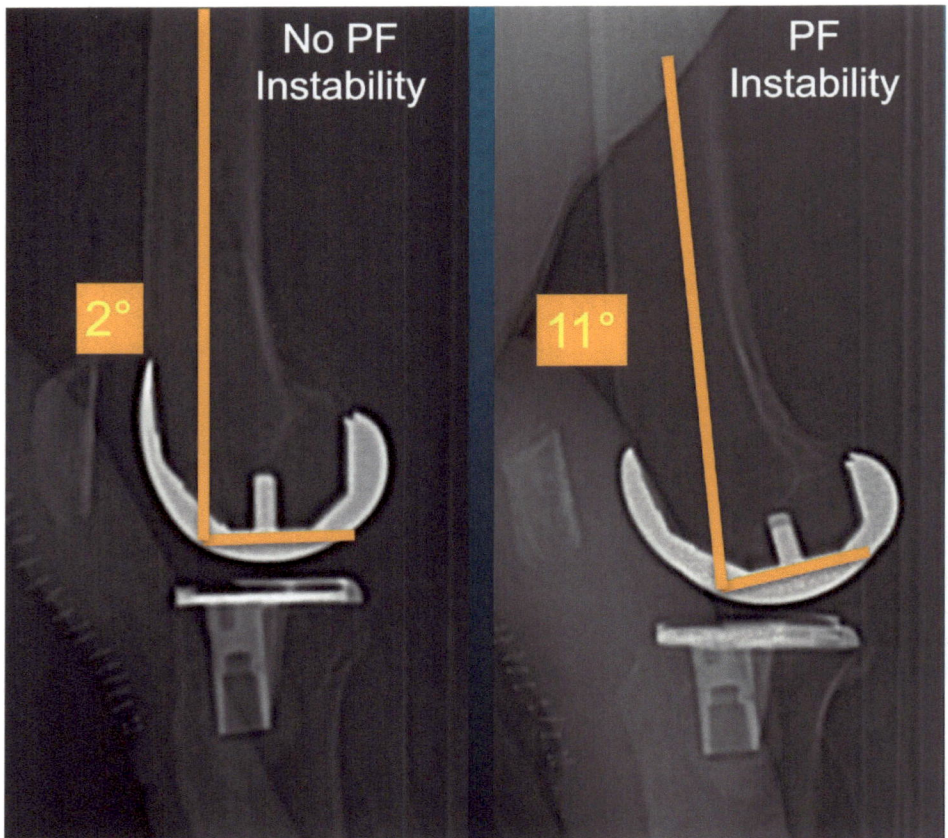

No PF Instability

2°

PF Instability

11°

This radiographic comparison shows a control patient (left) with 2 degrees of flexion of the femoral and a patient with patellofemoral instability (right) with 11 degrees of flexion. Placement in excessive flexion resulted from using a first-generation manual instrument and the lack of an intraoperative verification check.

This rudimentary instrumentation required a too posterior position for the starting hole for the intraosseous rod, which sets the distal cutting block in excessive flexion. The use of advanced manual instrumentation and a later-described verification check markedly reduce the risk of flexion.

PF Instability Occurred in Those Patients with Dome Patella

Speculate that soft-tissue overgrowth of the dome implant reduced congruency at 5 months

The use of a dome-shaped patella implant that is smaller and less congruent than the larger anatomic patella implant increased the risk of patellofemoral instability caused by a gradual soft-tissue overgrowth of the button over five months.

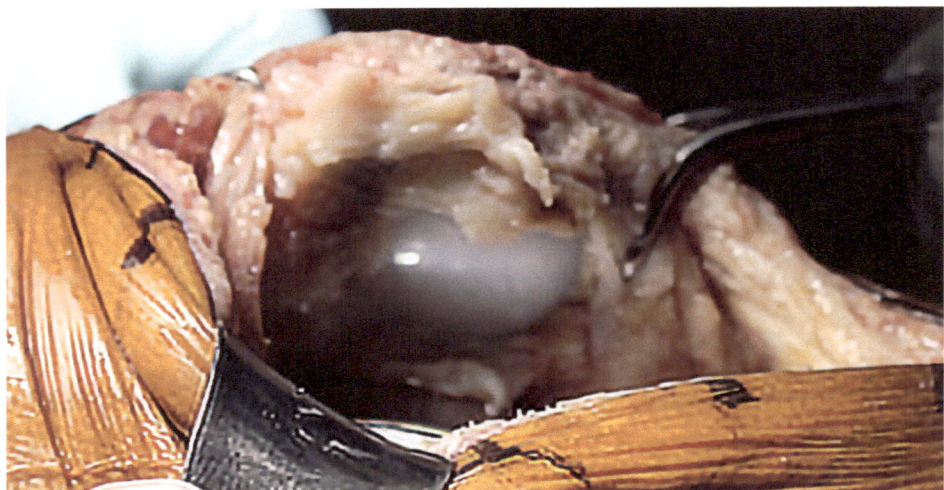

At the time of the patellofemoral stabilizing procedure, this medial view inside the right knee shows the dome-shaped patella implant is nearly half-covered with soft-tissue overgrowth.

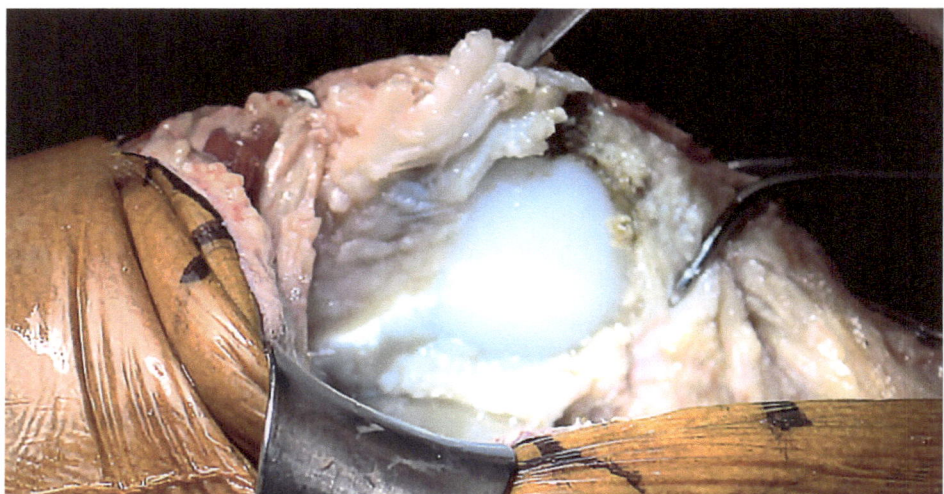

This view shows the soft-tissue overgrowth's partial dissection.

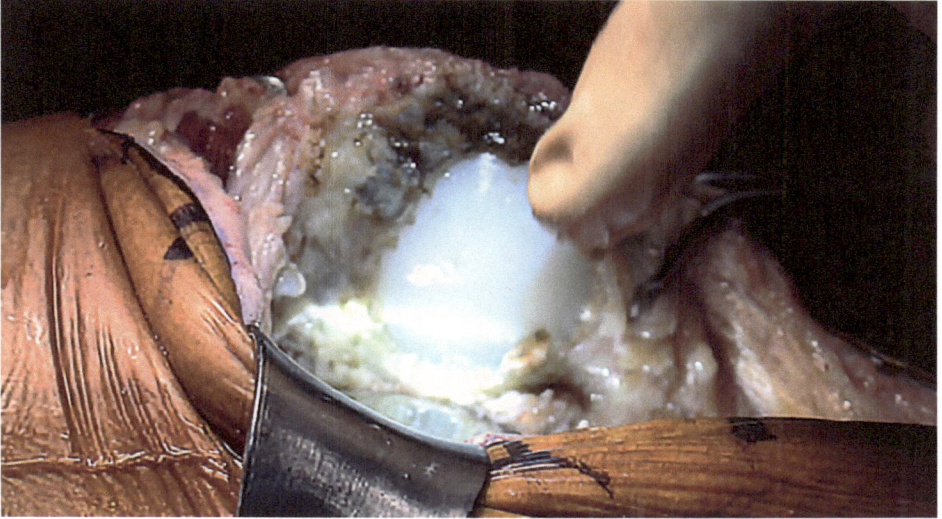

Finally, complete removal of the soft tissue overgrowth that accumulated over five months restores the dome-shaped patella's full convexity. It re-establishes the necessary congruency to contain the patella within the trochlear groove during activities.

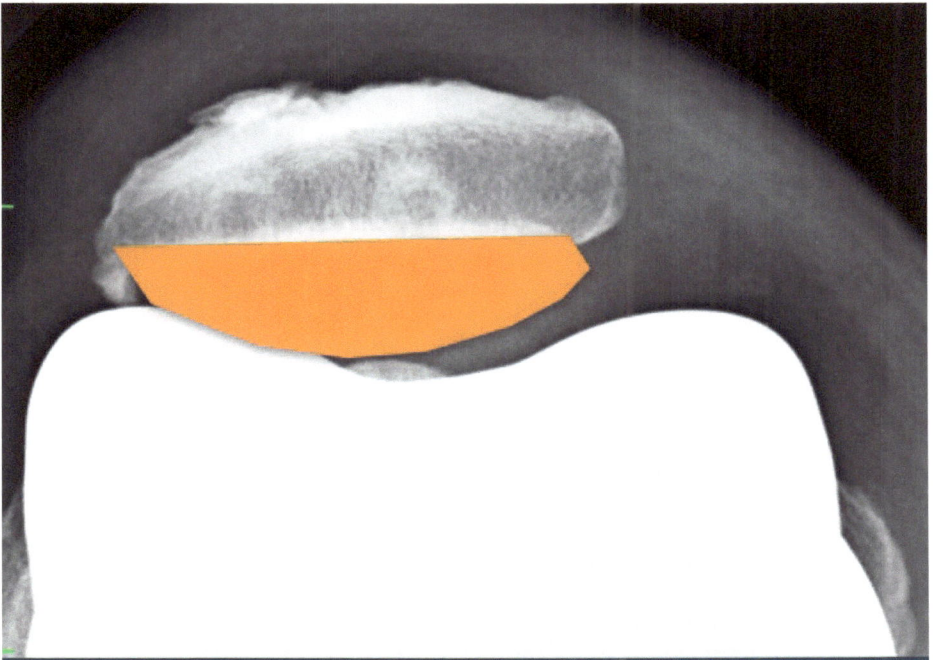

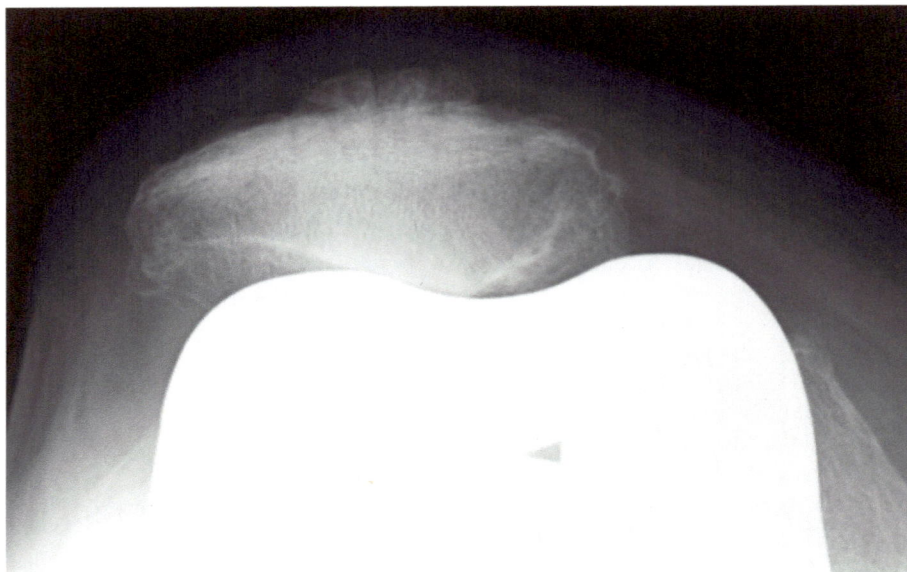

Patellofemoral instability did not occur in those patients with a large and congruent anatomic-shaped button that covered the patella's entire surface. It also did not occur in those patients with an un-resurfaced patella; however, 10% of these patients expressed dissatisfaction because of anterior knee pain. The current recommendation is to resurface the patella and do so with an anatomic shaped button.

Careful Positioning of the Entrance of the Drill Hole Reduces the Risk of Flexing the Femoral Component

The verification check that reduces the risk of flexing the femoral component is the careful positioning of the drill hole entrance in the distal femur that orients the intraosseous rod that sets the flexion of the distal femoral resection guide.

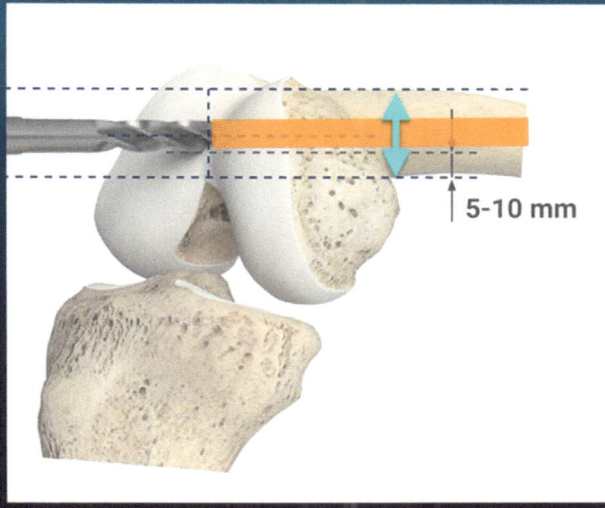

Center the drill hole midway between the anterior femoral cortex and top of notch

5-10 mm

This schematic shows the anatomic rationale for implementing the verification check to reduce the risk of flexing the femoral component. Notice that the femoral cortex's anterior and posterior aspects intersect the anterior half of the distal femoral condyles and not the posterior half.

Anatomically, this means that the drill hole needs to be centered in the anterior half of the distal femoral condyles to reduce the risk of flexing the femoral condyle. The surgeon should start the drill hole for the intraosseous rod, keeping a 5-10 mm bridge of bone between the hole's posterior edge and the intercondylar notch's anterior edge after removing osteophytes.

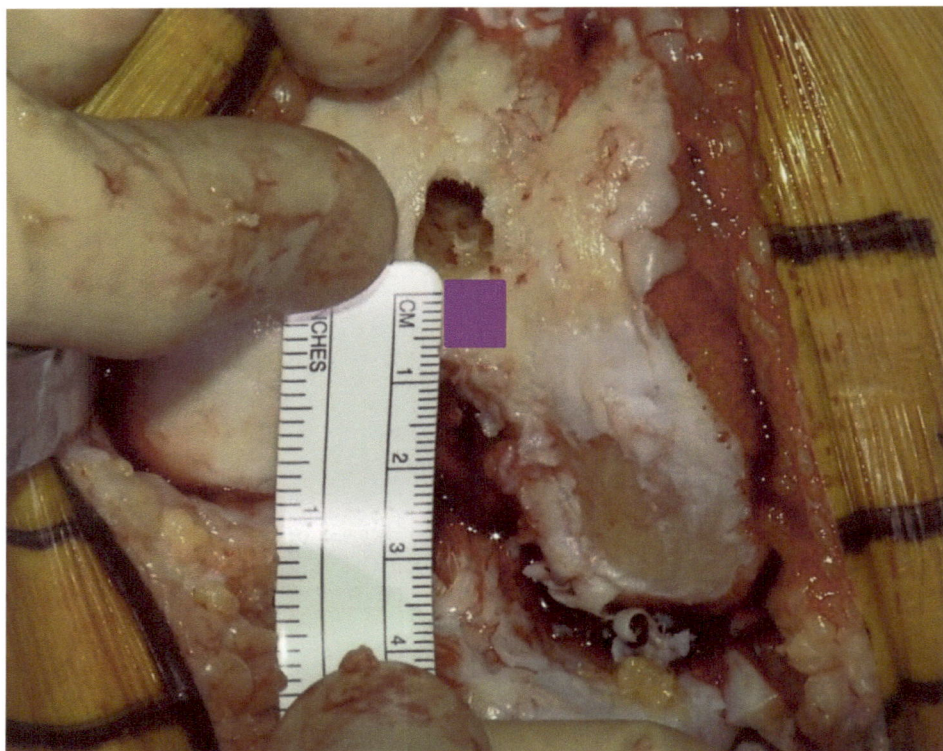

This intraoperative photograph of a right knee shows a ruler measuring the bone bridge, which is 6 mm and within the target of 5 to 10 millimeters.

Does a positioning rod or a patient-specific guide result in more natural femoral flexion in the concept of kinematically aligned total knee arthroplasty?

Max Ettinger[1] · Tilman Calliess[1] · Stephen M. Howell[2]

The success of the verification check's limiting flexion of the femoral component was reported by a study that compared the check of leaving a 5-10 mm bone bridge with the positioning the rod technique and the use of patient-specific guides that tend to rock a bit into flexion when seated. The bone bridge verification check resulted in a 1 degree mean flexion of the femoral component with little variability (standard deviation ± 2 degrees) and less flexion than the commonly used patient-specific guides.

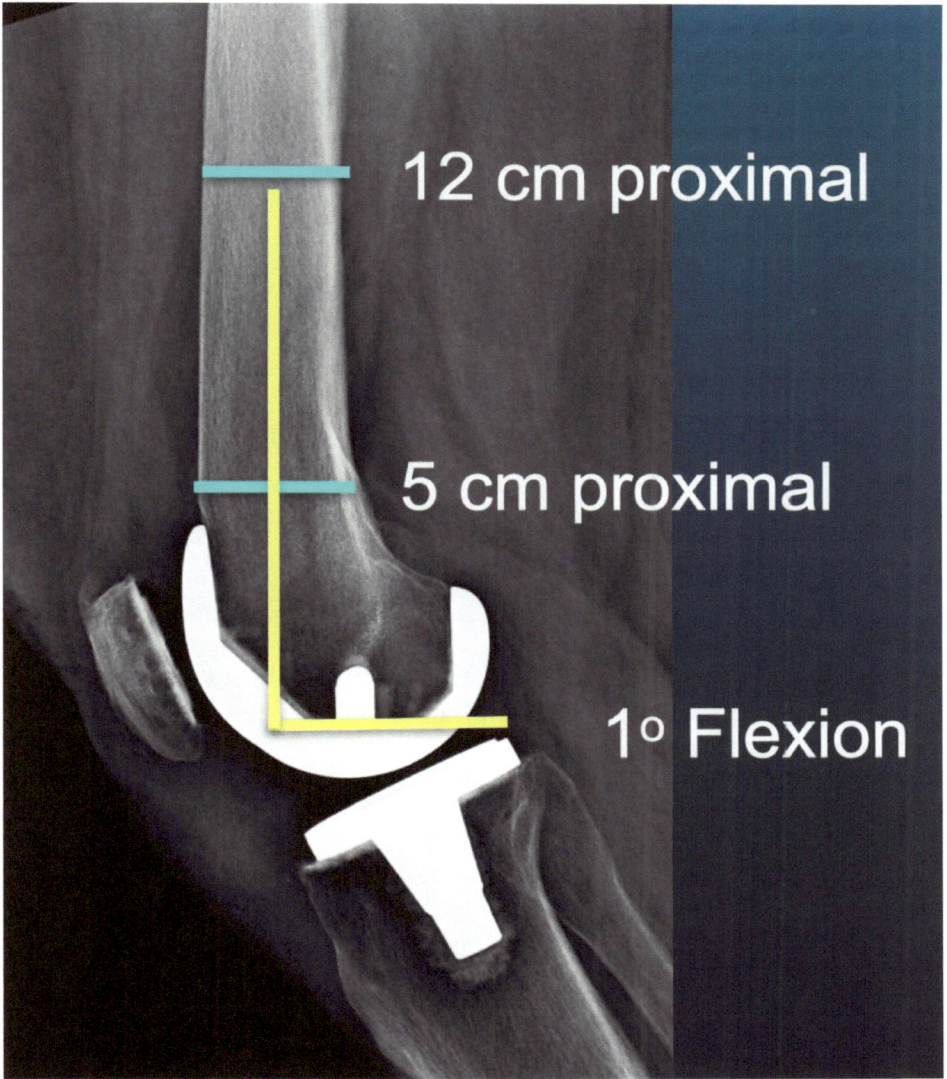

The measurement method detailed on a lateral radiograph of a representative patient shows one degree of flexion of the femoral component, which is within the target.

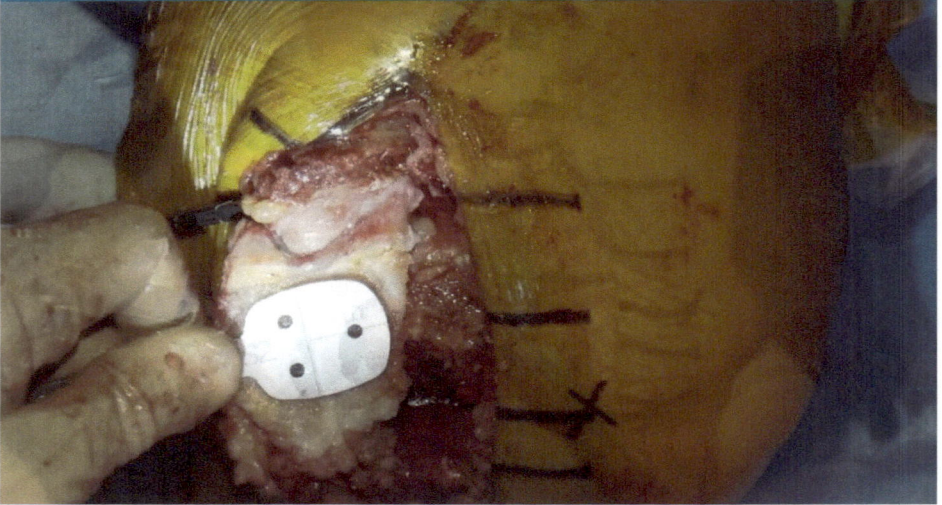

Use an Anatomic and Not a Dome Patella component to Maximally Cover Patella

The intraoperative photograph illustrates the drill holes' placement for an anatomic-shaped patella implant to maximize medial-lateral coverage of the patella resection.

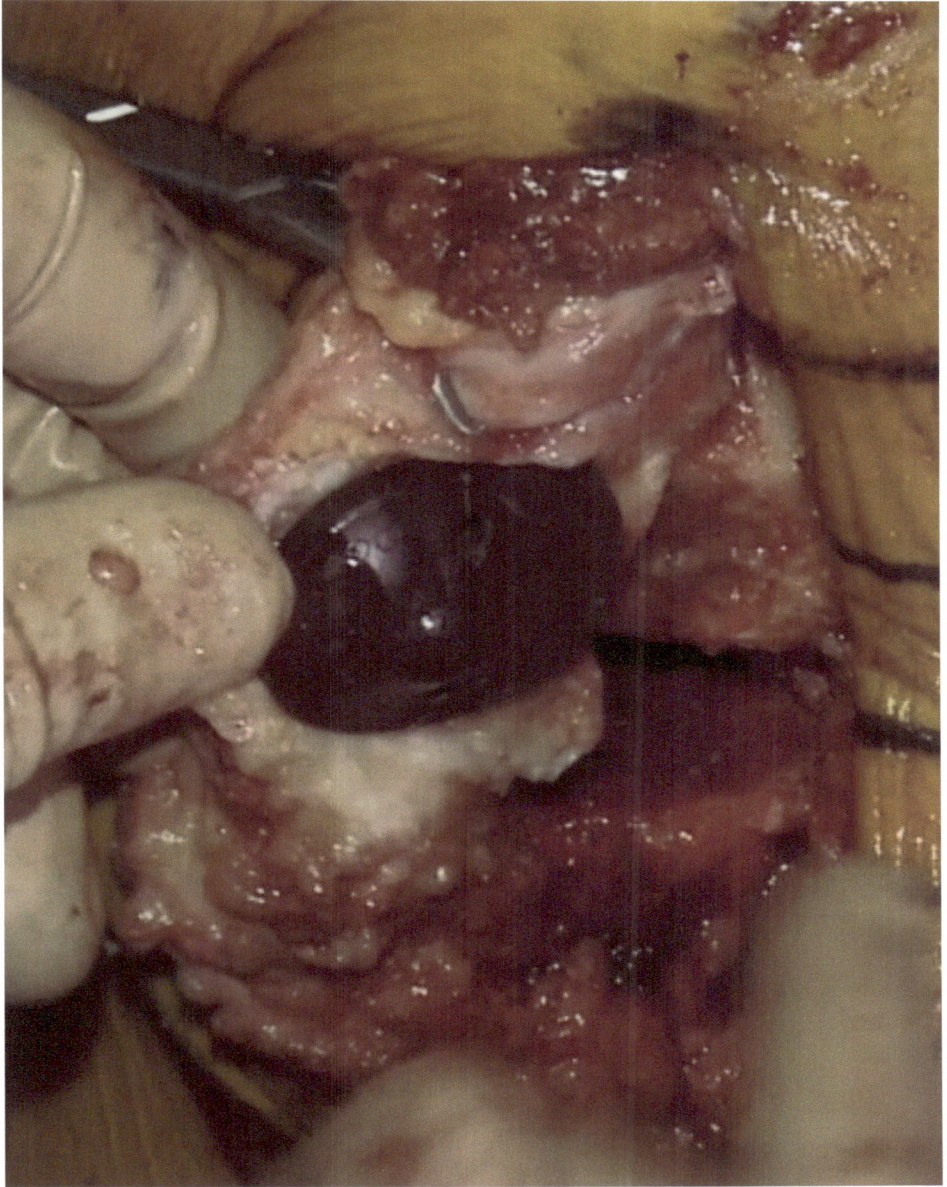

Compared to the dome-shaped patella implant, the larger anatomic-shaped implant provides more contour to stay centered in the groove during flexion, more coverage of the patella resection, and less risk of extensive soft-tissue overgrowth.

Consider following the 8th Commandment:

8.   THOU SHALT NOT FLEX THE FEMORAL COMPONENT MORE THAN A FEW DEGREES OR RISK PATELLOFEMORAL INSTABILITY.

Chapter 9 The 9th Commandment

The 9th commandment is **Thou Shalt Not Deviate from the Patient's Pre-Arthritic Slope More Than a Few Degrees or Risk Early-Onset of Tibial Component Failure from Posterior Overload.**

`

9.   THOU SHALT NOT DEVIATE FROM THE PATIENT'S PRE-ARTHRITIC SLOPE MORE THAN A FEW DEGREES OR RISK EARLY-ONSET TIBIAL COMPONENT FAILURE FROM POSTERIOR OVERLOAD.

The following large series and case-matched study identified an association between setting the tibial component's posterior slope greater than the patient's pre-arthritic slope and early-onset tibial component failure.

What mechanisms are associated with tibial component failure after kinematically-aligned total knee arthroplasty?

Alexander J. Nedopil[1] ⓘ · Stephen M. Howell[2] · Maury L. Hull[3]

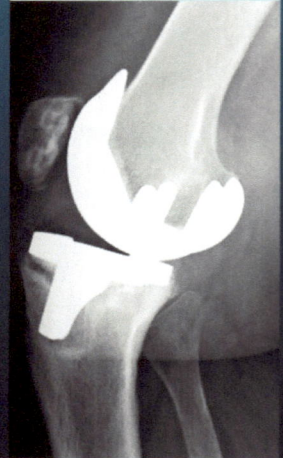

- Only 0.3% tibial components failed (8 of 2725; 2-9 year f/u)

Posterior Subsidence of Tibial Component

Nedopil, Int Orthop, 2017

A prospective patient database review identified 8 out of 2,725 consecutive knees with early tibial component failure with a 2 to 9 year follow up. This proportion equates to a 0.3% incidence, which is 3-4 times lower than the 1 to 1.5% incidence of tibial component failure reported by comparable studies of mechanically aligned TKA.

The mechanism of failure after calipered kinematic alignment was not varus overload like mechanical alignment, but posterior subsidence or posterior rim wear of the insert. The mean time of presentation was 28 months post-operatively, with the longest at five years.

The following series of radiographs illustrate the mechanism of posterior tibial component failure using a representative patient.

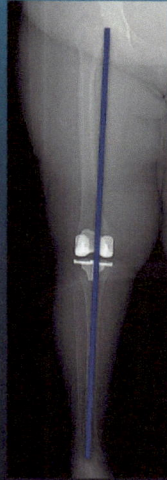

Mechanisms of Failure Were Posterior Loosening and Posterior Insert Wear

Post Op

- Five patients had posterior tibial loosening

Nedopil, Int Orthop, 2017

The limb alignment of the patient after calipered kinematic alignment met the target for mechanical alignment target, so an error in the component alignment in the coronal plane does not explain the etiology of failure.

The component orientation met the target for kinematic alignment in the coronal plane as caliper measurements of the bone resections and the use of the serial verification checks restored the patient's pre-arthritic joint lines. Hence, the cause of failure is more easily explained by an error in the sagittal plane component position and not the coronal plane.

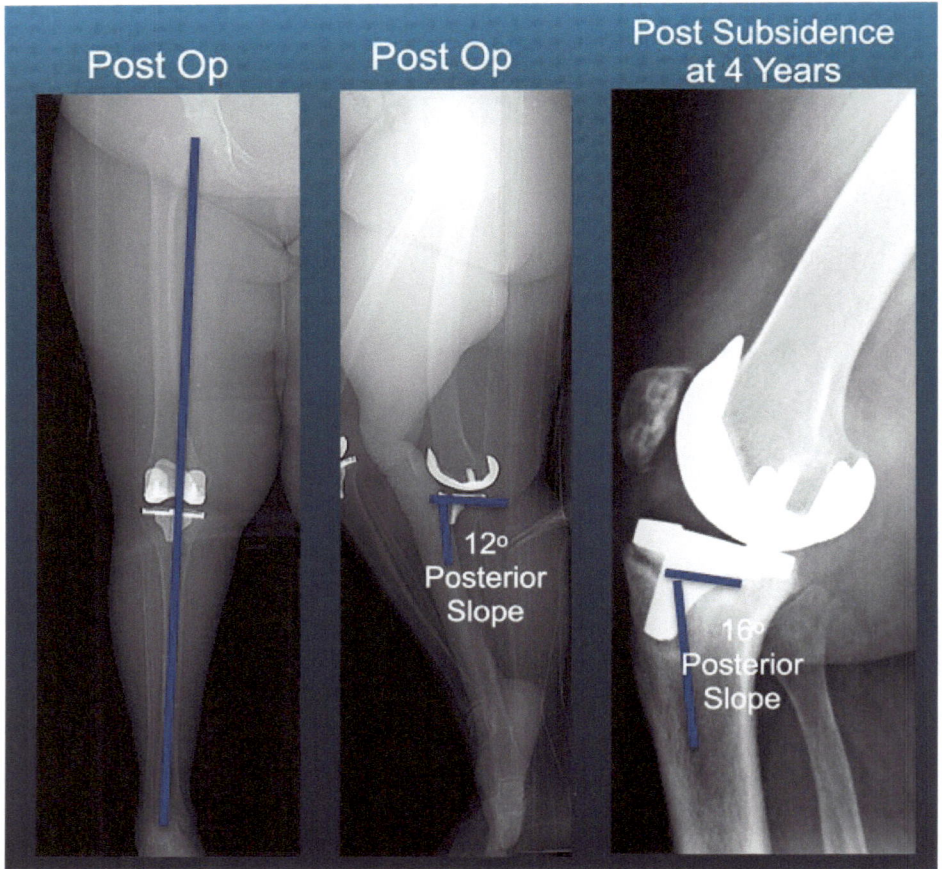

| Post Op | Post Op | Post Subsidence at 4 Years |

12º Posterior Slope

16º Posterior Slope

The day after the operation, these serial post-op lateral radiographs show the tibial component in 120 posterior tibial slope that is 60 greater than the pre-arthritic or pre-operative slope.

Three years later, the patient presented with effusions, pain, and instability from posterior subsidence of the tibial component due to a 40 increase in posterior slope to 160.

Mechanistically, when the slope exceeds the pre-arthritic slope, the flexion space is too loose, enabling the femoral condyles to contact and load the posterior half of the base plate and insert daily living activities.

In the case series, posterior over-load caused posterior subsidence and anterior elevation tibial baseplate in five patients, and posterior rim wear of the insert in three.

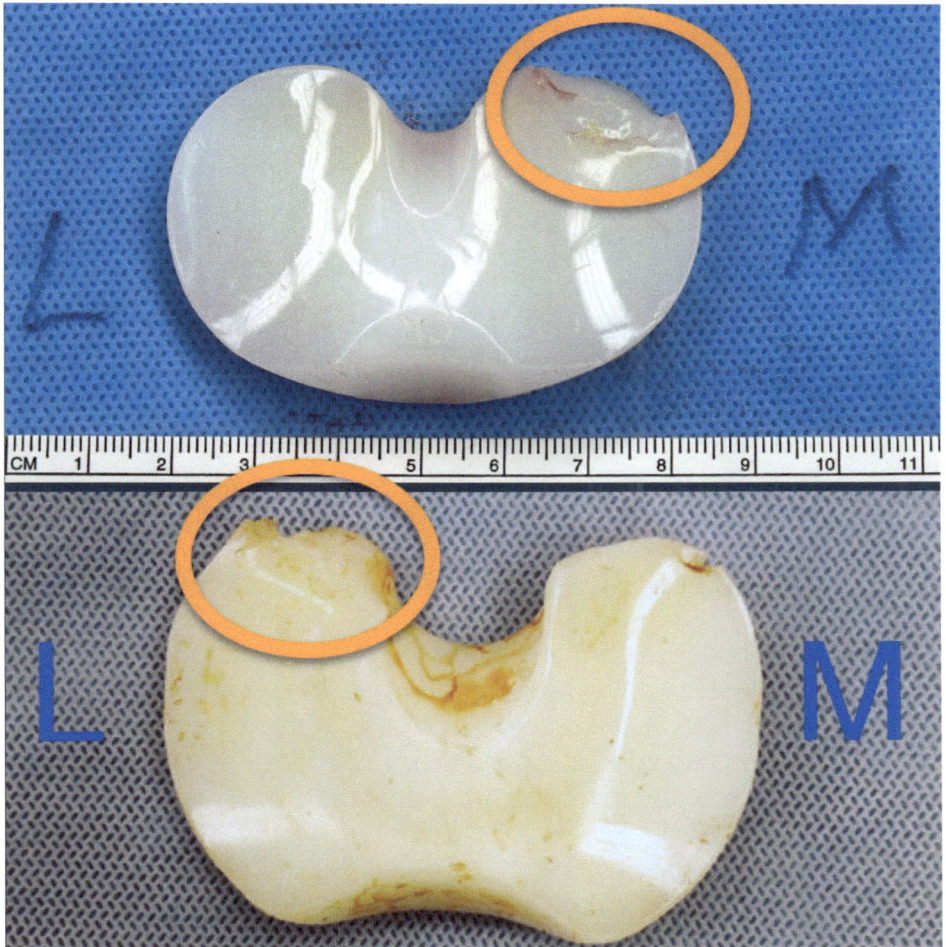

These photographs of a tibial insert retrieved at the time of revision surgery show posteromedial rim wear (above) and posterolateral rim wear (below). Patients with poly wear present with increasing pain, effusion, and instability over five to six months. The tibia begins to spin-out while walking and turning like the patient with rotatory instability with a posterior horn meniscectomy and an ACL tear. Treatment of the tibial component's posterior failure after calipered kinematically aligned TKA is straight-forward; revise the tibial component, reduce the slope, and put it in a new insert.

Setting Slope of Tibial Component (TC) More Posterior Than Native Caused Failure

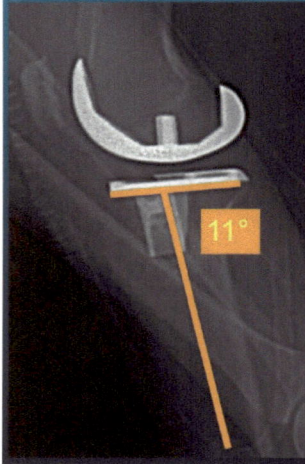

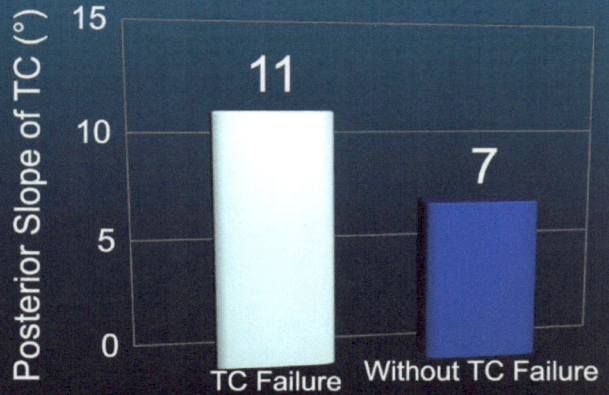

The case-matched series showed the need to closely match the pre-arthritic slope when using an implant design that retains the posterior cruciate ligament. A mean 4 degree increase in the posterior slope caused a posterior overload and tibial component failure when using a posterior cruciate ligament retaining implant design.

With posterior cruciate ligament excision, there might be a need to reduce the pre-arthritic to compensate for the increase in flexion space laxity and reduce the risk of using a thicker insert that compromises knee extension.

The following is a visual verification check that reduces the risk of deviating from the pre-arthritic slope when setting up the tibial resection.

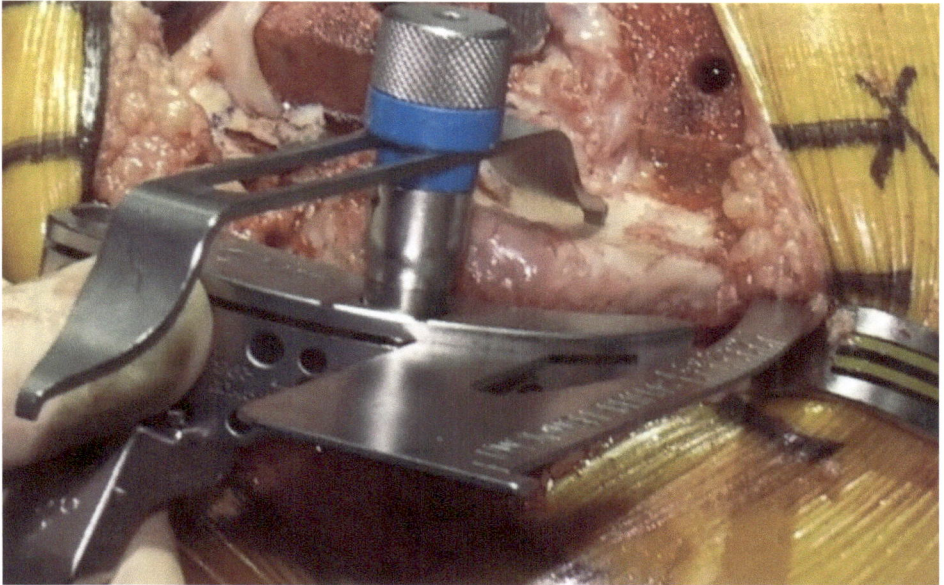

Exposure of the posteromedial tibia and an angel wing's placement in the medial side of the saw slot enables the surgeon to see the tibial resection's flexion-resection plane relative to the medial tibial articular surface.

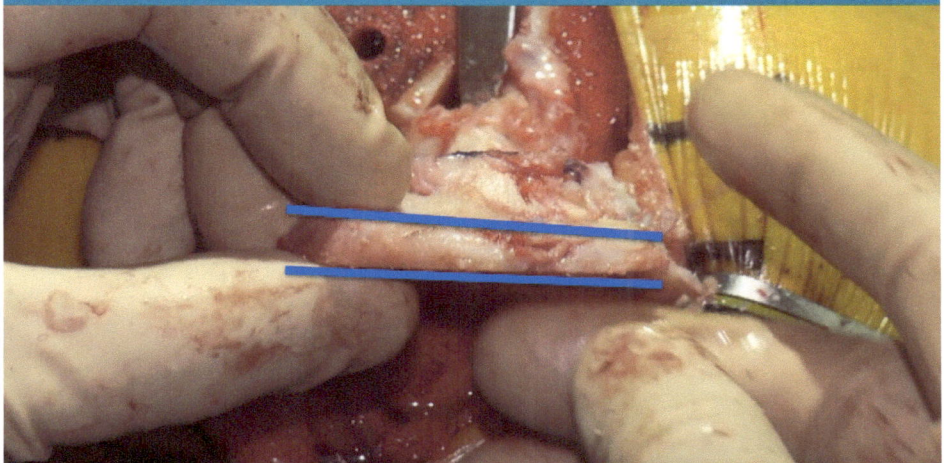

Verification Reduces Risk of Not Matching Pre-Arthritic Slope & Posterior Loosening/Wear

Check Angel Wing in Saw Slot, Adjust Resection to Match Pre-Arthritic Slope, Inspect Slope of Resection, Fine-Tune Slope

The surgeon visually compares the degree of parallelism between the resection plane and the tibial articular surface. This photograph of the medial side of the tibial resection of a typical varus deformity with medial wear shows the thickness of the anterior edge (held by the gloved thumb and index finger left) is a little thicker than the posterior edge.

Intentionally starting the resection with a few degrees less posterior slope than the pre-arthritic one is appropriate, especially when using an implant design that requires resection of the posterior cruciate ligament. Adjustment of the slope to reduce tightness and laxity in flexion reduces the tibial component's risk of posterior subsidences.

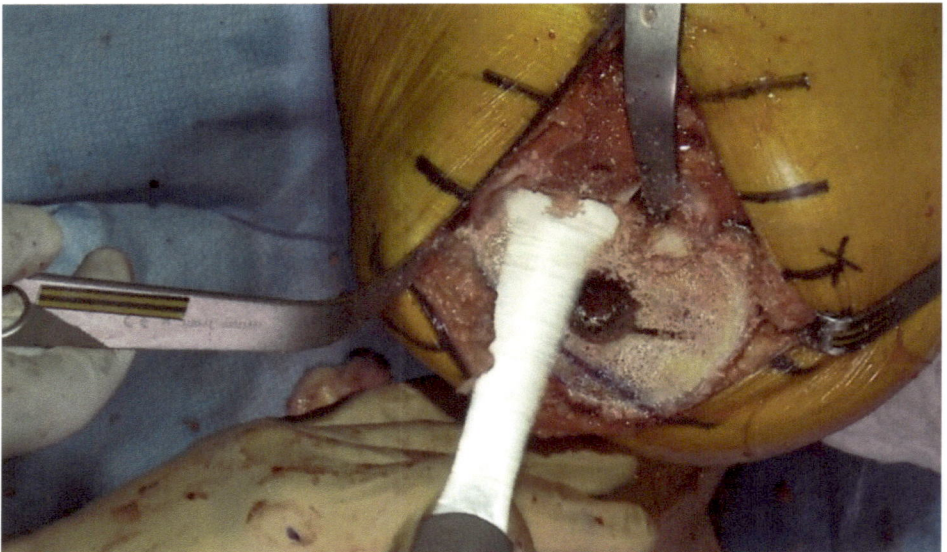

The photograph of the top view of the tibial resection (medial left) shows the use of a saw blade to 'feather' 1-2 millimeters of bone from the posterior tibia to increase the slope a few degrees using the 1.2 mm thickness of the saw blade as a gauge for the amount of bone removal.

Consider following the 9th Commandment:

9.   THOU SHALT NOT DEVIATE FROM THE PATIENT'S PRE-ARTHRITIC SLOPE MORE THAN A FEW DEGREES OR RISK EARLY-ONSET TIBIAL COMPONENT FAILURE FROM POSTERIOR OVERLOAD.

Chapter 10 The 10th Commandment

The 10th Commandment of calipered kinematically aligned total knee arthroplasty is **Thou Shalt Not Intentionally Remove the PCL, Though When Cut Reduce the Posterior Tibial Slope to Compensate for the Increase in Laxity in the Flexion Space.**

10. THOU SHALT NOT INTENTIONALLY REMOVE THE PCL, THOUGH WHEN CUT, REDUCE THE POSTERIOR TIBIAL SLOPE TO COMPENSATE FOR THE INCREASE IN LAXITY IN THE FLEXION SPACE.

A mid-section resection of the PCL occurs when the blade skips towards the center of the knee when doing the posterior chamfer cut. An insertional resection of the PCL occurs when cutting the tibia from either cutting too much slope and too much tibia.

Excision of the PCL creates an unpredictable and variable increase in flexion space laxity and a change from a natural trapezoidal to a rectangular shape.

■ KNEE

Posterior cruciate ligament resection in total knee arthroplasty

THE EFFECT ON FLEXION-EXTENSION GAPS, MEDIOLATERAL LAXITY, AND FIXED FLEXION DEFORMITY

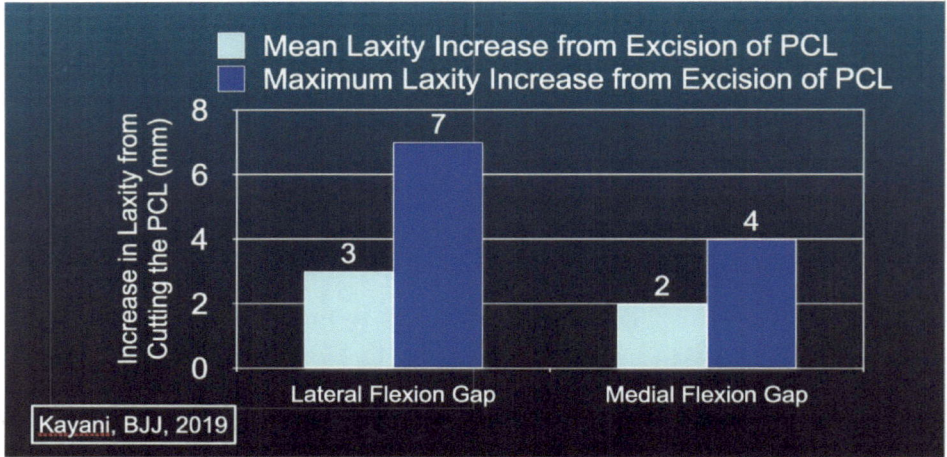

Kayani, BJJ, 2019

The column graph shows the substantial and variable increase in the column graph shows the substantial and variable increase in the medial and lateral gaps or compartments in 90 degrees of flexion from cutting the PCL at the time of total knee arthroplasty as measured intraoperatively with robotics.

Cutting the PCL causes a mean 3 mm increase in the lateral flexion gap, and the maximum, when you factor in the two standard deviations, the increase becomes 7 mm. Cutting the PCL causes a mean 2 mm increase in the medial flexion gap, and the maximum, when you factor in the two standard deviations, the increase becomes 4 mm.

The PCL's medial and non-central location explains the change in the flexion space's shape from trapezoidal to rectangular from PCL resection.

The magnitude of the increase in flexion space laxity is not identifiable pre-resection of the PCL. When an unexpectedly large laxity increase occurs, there is a risk that constrained implants are required, and the surgeon may be unprepared.

Calipered kinematic alignment retains the PCL to maintain the patient's pre-arthritic medial and lateral flexion gaps and trapezoidal flexion space, which reduces the need for added constraint. Retaining the PCL results in the patient sensing the TKA feels like a normal knee.

Several methods try to compensate for the unpredictable laxity increase in the flexion space from resection of the PCL, understanding they don't fully correct the abnormal laxity.

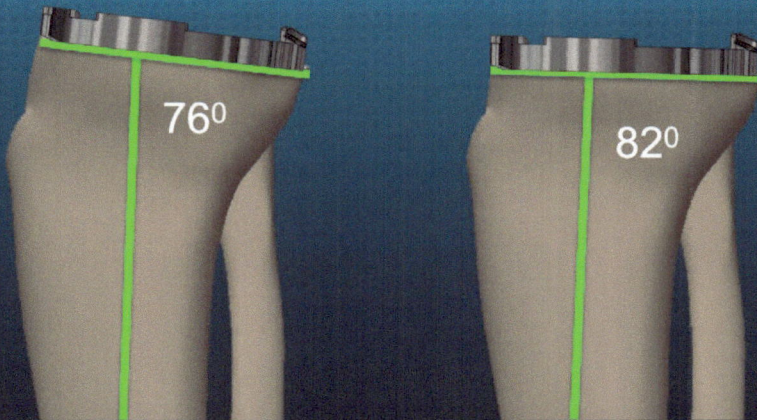

The schematic shows the re-cut method to reduce the tibial component's posterior slope, which lowered the slope from 76 to 82 degrees. Increasing the insert thickness tightens the flexion space with a smaller effect on tightening the extension space. This technique is preferred as long as the insert's final thickness after reduction of the slope is 14 mm or less.

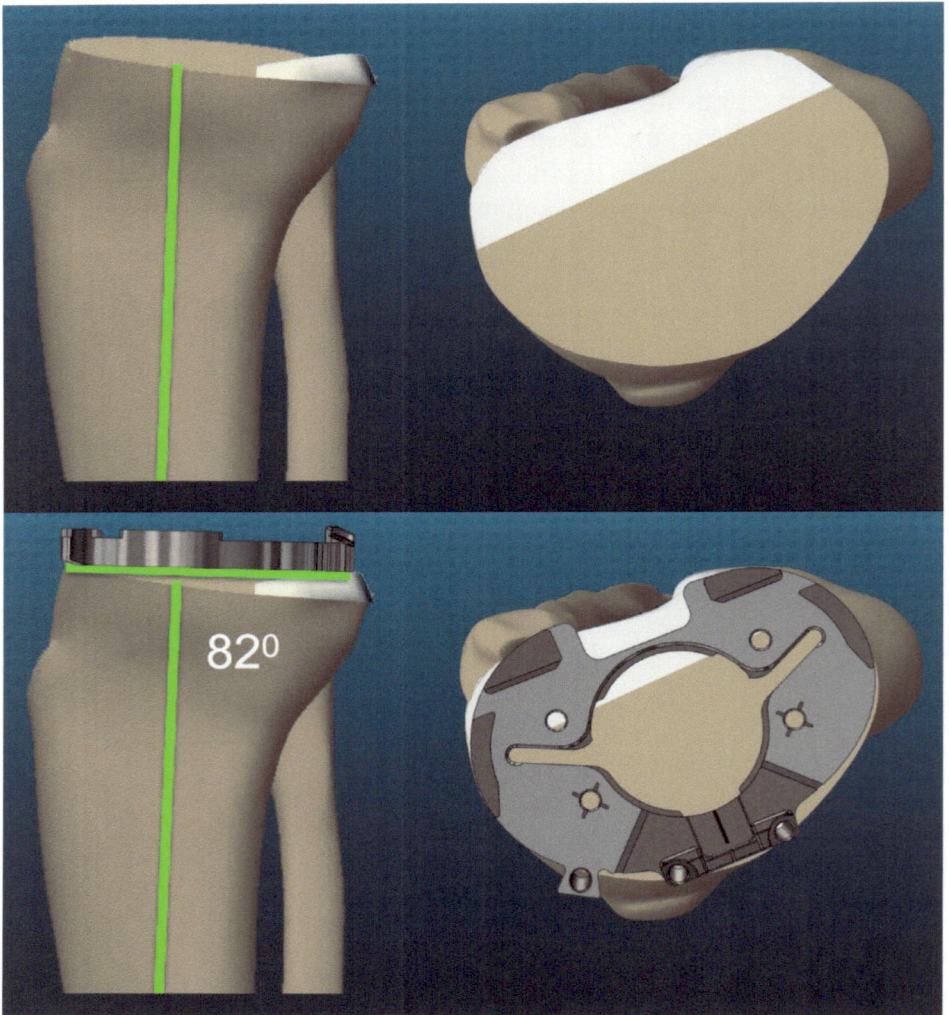

The schematic shows the posterior bone grafting method to reduce the tibial component's posterior slope, which lowered the slope to 82 degrees. A saw prepares the bone graft by cutting a wedge off the excised proximal tibia. Posterior tibial bone grafting is preferred when the final insert thickness before reducing the slope was 13 mm or greater.

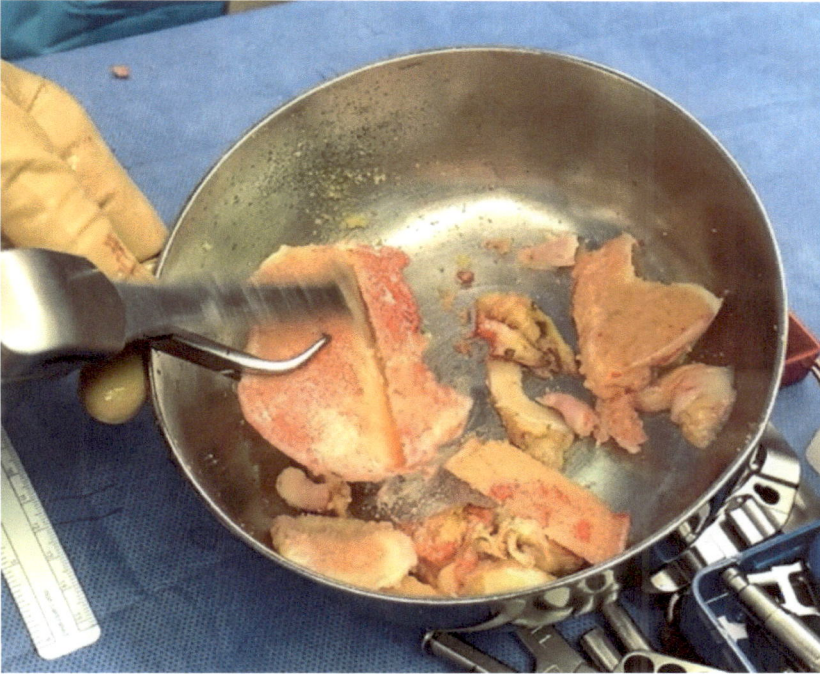

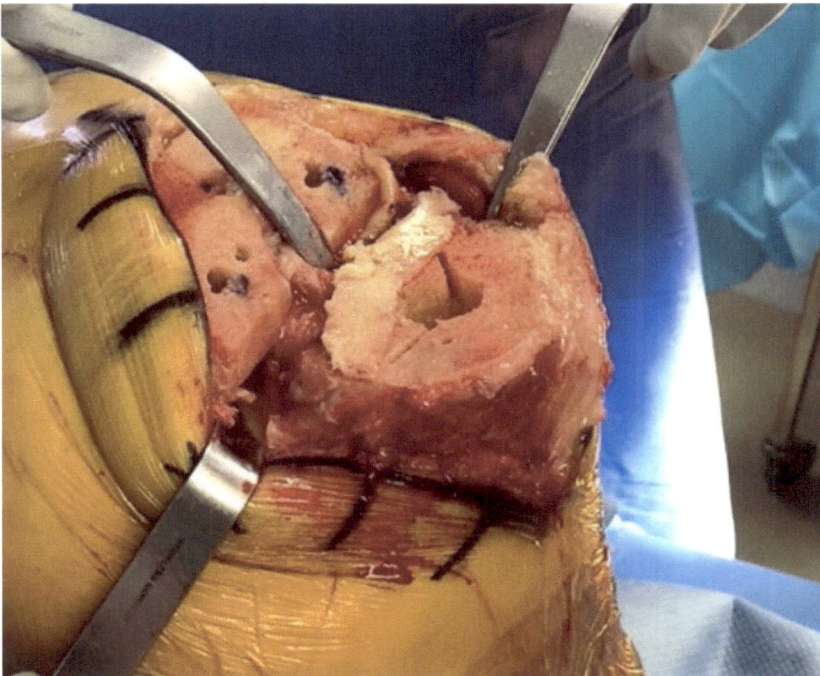

The intraoperative photographs show a saw removing a wedge of bone, tapered to 5 mm thick, from the posterior portion of the excised proximal tibia to make the bone graft (left). Placement of the bone graft on the posterior third of the proximal tibia reduces the slope (right).

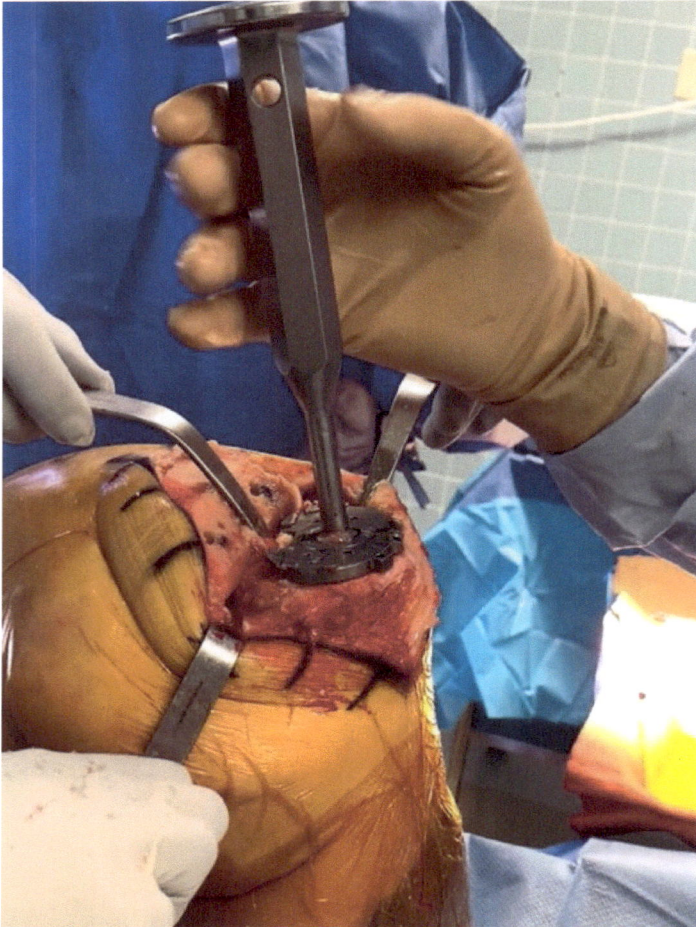

A mallet drives the cruciate stem punch to compact the bone graft and set the tibial baseplate's desired slope level.

Bone Graft Restores Patient's Pre-Arthritic Flexion Space A-P Stability

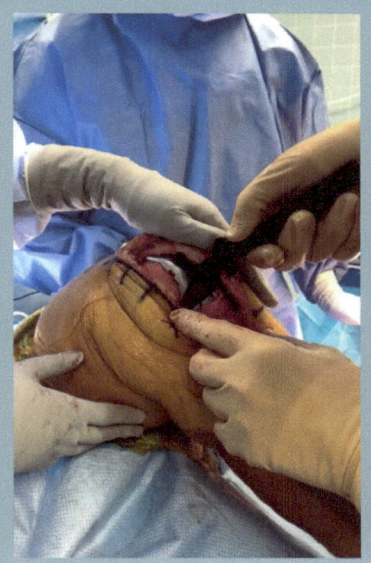

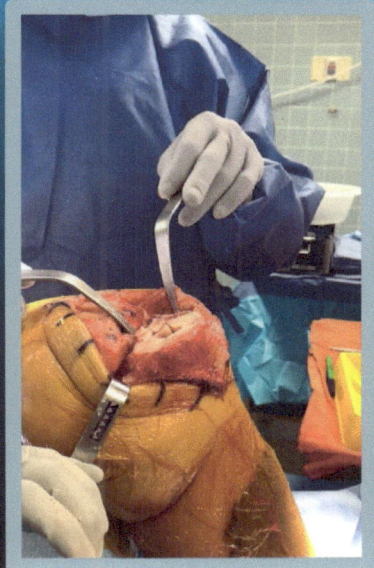

The surgeon performs a trial reduction and verifies that the posterior slope reduction is correct (left). With the knee in 90 degrees of flexion, the passive internal-external rotation should be ~20 degrees. The anterior offset of the tibia referenced to the distal medial femur should match the value at exposure when measured with an offset caliper. The bone graft grips the tibial resection like Velcro so that it remains interdigitated after pressure washing with a water-pick (right).

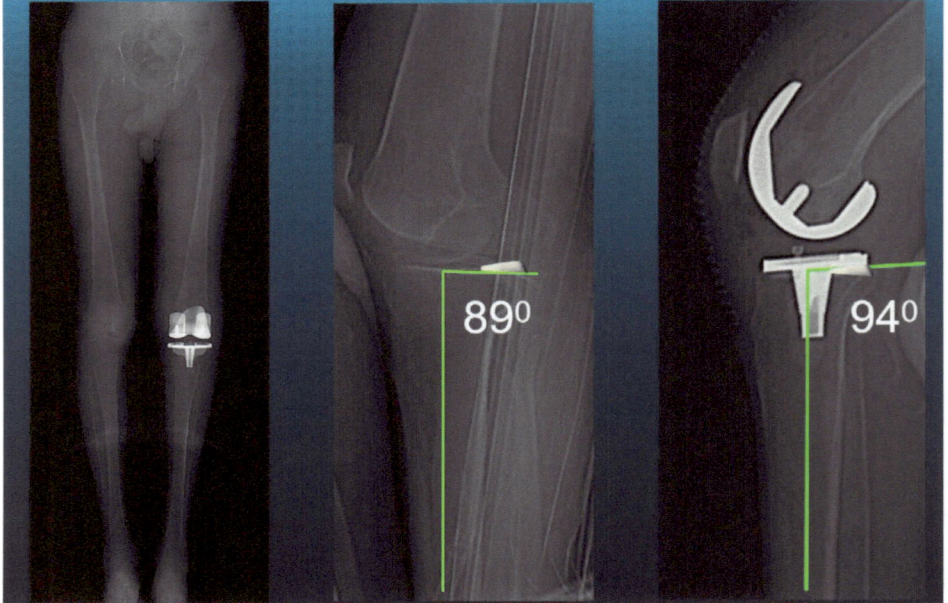

Post-Op Radiographs Show Posterior Tibial Bone Graft Reduced Patient's Pre-Arthritic Posterior Slope

89⁰

94⁰

Post-operative radiograph shows the posterior bone graft reduced the patient's pre-arthritic slope from 89 degrees (center image) to 94 degrees (right image), which lessened the increase in flexion space laxity caused by resection of the PCL.

This concludes the 10th Commandment:

10. THOU SHALT NOT INTENTIONALLY REMOVE THE PCL, THOUGH WHEN CUT, REDUCE THE POSTERIOR TIBIAL SLOPE TO COMPENSATE FOR THE INCREASE IN LAXITY IN THE FLEXION SPACE.

Chapter 11 Not a Commandment— Choosing Your Orthopedic Surgeon

When you meet your orthopedic surgeon, recognize that they spent many years of academic commitment and dedication learning their craft. Each has a four-year undergraduate degree, followed by four years of medical school, five years of orthopedic residency, and often an additional year of fellowship training before finally 'starting' practice in their mid-30s. The orthopedic surgeon makes this commitment as there is immense satisfaction from improving a patient's quality of life and the heartfelt testimonials from patients and families.

Despite these training similarities, there are differences in how each orthopedic surgeon chooses to practice. From the authors' viewpoint, the alignment method is the most important predictor of your satisfaction and function after a total knee replacement. Combined, we spent the first 40 years of our career practicing mechanical alignment and switched to calipered kinematic alignment for the last 30 years.

After using most of the available instrument systems for positioning the components, different component designs, and newer technologies, it became clear that each performed better when calipered kinematic alignment was used instead of mechanical alignment. Simply put, the impact of alignment overrides all other issues in determining how quickly you recover and how much you will like your total knee replacement.

When you are looking for an orthopedic surgeon to perform your knee replacement, recognize that you will likely forge a permanent relationship with them. Let them know what type of alignment you would like and gauge their experience. When you choose kinematic alignment, ask them if they measure the thickness of the four bone resections from the femur and the tibial resection thickness with a caliper. Ask them whether they correct bone resections thicker or thinner by 1 mm from the kinematic alignment target. They need to commit to using a caliper and applying corrections even when using robotics, navigation, and patient-specific instrumentation, as there are inherent imperfections in all instrument systems.

You can assess the commitment to caliper kinematic alignment from conversations with the surgeon and their office and hospital team. Study their handouts and videos and attend the pre-surgery class at the hospital or same-day surgery center. When you remain uncertain, ask to attend the pre-surgery class at the hospital or surgical center BEFORE making your decision. Recognize that in 2020, most patients having total knee replacement can expect a same-day discharge or one-night hospital stay with short-term use of low-dose narcotics.

Once you have your total knee replacement, please share what you like and don't like about it with your surgeon. They continually strive to improve the patient experience and can only do so with your comments and critique!

Other Books by Authors / Editors (amazon.com, and Kindle)

The Medical Advisor by Thomas D Meade MD, ICS BOOKS, INC
OTISMED-ShapeMatch Technology, Kinematic Alignment Reference Guide by
Thomas D Meade MD, Stephen M Howell MD, Stryker Publication

Christmas Wings for Brian A heartwarming story of a boy whose shoulders kept growing
Merry Christmas to Wilkes-Barre 50 Ways" for Mayor George Brown to Create a Better City.
Air Force Football Championship Seasons from AF Championship to Coach Calhoun's latest team
Syracuse Football Championship Seasons Beginning of S.U. Football championships; to Dino Babers' Era
Navy Football Championship Seasons 1ˢᵗ Navy Championships to the Ken Niumatalolo **Era**
Army Football Championship Seasons Beginning of Football to the championships of Jeff Monken Era
Florida Gators Championship Seasons Beginning of Football to championships of Dan Mullen era
Alabama's Championship Seasons Beginning of Football past the 2017/2018 National Championship
Clemson Tigers Championship Seasons Beginning of Football to the 2018/2019 Clemson National Champs
Penn State's Championship Seasons PSU's first championship to the James Franklin era
Notre Dame's Championship Seasons Before Knute Rockne and pastLou Holtz's 1988 undisputed title
Super Bowls & Championship Seasons: The New York Giants Many championship seasons of the Giants.
Super Bowls & Championship Seasons: New England Patriots Many championship seasons of the Patriots.
Super Bowls & Championship Seasons: The Pittsburgh Steelers Many championship seasons of the Steelers
Super Bowls & Championship Seasons: The Philadelphia Eagles Many seasons of the Eagles.
The Big Toxic School Wilkes-Barre Area's Tale of Corruption, Deception, Taxation & Tyranny
Great Players in New York Giants Football Begins with great players of 1925 to the Saquon Barqley era.
Great Coaches in New York Giants Football Begins with Bob Folwell in 1925 and to Pat Shurmur in 2019.
Great Moments in New York Giants Football Beginning of Football to the Pat Shurmur era.
Hasta La Vista California Give California its independence.
I.T.'s ALL OVER! Mueller: "NO COLLUSION!"—Top Dems going to jail for the hoax!
Democrat Secret for Power & Winning Elections Open borders adds millions of new Democrat Voters
Hope for Wilkes-Barre—John Q. Doe—Next Mayor of Wilkes-Barre
The John Doe Plan & W.B. Plan will help create a better city!
Great Moments in New England Patriots Football Second Edition
This book begins at the beginning of Football and goes to the Bill Belichick era.
The Cowardly Congress Corrupt U.S. Congress is against America and Americans.
Great Players in Air Force Football From the beginning to the current season
Great Coaches in Air Force Football Grom the beginning to Coach Troy Calhoun
Help for Mayor George and Next Mayor of Wilkes-Barre How to vote for the next Mayor Council
Ghost of Wilkes-Barre Future: Spirit's advice for residents about how to pick the next Mayor and Council
Great Players in Air Force Football: Air Force's best players of all time
Great Coaches in Air Force Football: From Coach 1 to Coach Troy Calhoun
Great Moments in Air Force Football: From day 1 to today
Great Players in Navy Football: Navy's best including Bellino & Staubach
Great Coaches in Navy Football: From Coach 1 to Coach #39 Ken Niumatalolo
Great Moments in Navy Football: From day 1 to coach Ken Niumatalolo 1
No Tree! No Toys! No Toot! Heartwarming story. Christmas gone while 19 month old napped
How to End DACA, Sanctuary Cities, & Resident Illegal Aliens . best solution to wipe shadows in America.
Government Must Stop Ripping Off Seniors' Social Security!: Hey buddy, seniors can't spare a dime?
Special Report: Solving America's Student Debt Crisis!: The only real solution to the $1.52 Trillion debt
The Winning Political Platform for America Unique winning approach to solve problems in America.
Lou Barletta v Bob Casey for U.S. Senate Barletta's unique approach to solving the big problems in America.
John Chrin v Matt Cartwright for Congress Chrin has a unique approach to solve big problems in America.
The Cure for Hate !!! Can the cure be any worse than this disease that is crippling America?
Andrew Cuomo's Time to Go? "He Was Never that Great!": Cuomo says America never that great
White People Are Bad! Bad! Bad! Whoever thought a popular slogan in 2018 would be *It's OK to be White!*
The Fake News Media Is Also Corrupt !!!: Fake press / media today is not worthy to be 4ᵗʰ Estate.
God Gave US Donald Trump? Trump was sent from God as the people's answer
Millennials Say America Was "Never That Great": Too many pleased days of political chumps not over!
It's Time for The John Q. Doe Party... Don't you think? By Elephants.
Great Players in Florida Gators Football... Tim Tebow and a ton of other great players
Great Coaches in Florida Gators Football... The best coaches in Gator history.
The Constitution by Hamilton, Jefferson, Madison, et al. The Real Constitution
The Constitution Companion. Will help you learn and understand the Constitution
Great Coaches in Clemson Football The best Clemson Coaches right to Dabo Swinney

Great Players in Clemson Football The best Clemson players in history
Winning Back America. America's been stolen and can be won back completely
The Founding of America… Great book to pick up a lot of great facts
Defeating America's Career Politicians. The scoundrels need to go.
Midnight Mass by Jack Lammers… You remember what it was like Great story
The Bike by Jack Lammers… Great heartwarming Story by Jack
Wipe Out All Student Loan Debt--Now! Watch the economy go boom!
No Free Lunch Pay Back Welfare! Why not pay it back?
Deport All Millennials Now!!! Why they deserve to be deported and/or saved
DELETE the EPA, Please! The worst decisions to hurt America
Taxation Without Representation 4th Edition Should we throw the TEA overboard again?
Four Great Political Essays by Thomas Dawson
Top Ten Political Books for 2018… Cliffnotes Version of 10 Political Books
Top Six Patriotic Books for 2018… Cliffnotes version of 6 Patriotic Boosk
Why Trump Got Elected!.. It's great to hear about a great milestone in America!
The Day the Free Press Died. Corrupt Press Lives on!
Solved (Immigration) The best solutions for 2018
Solved II (Obamacare, Social Security, Student Debt) Check it out; They're solved.
Great Moments in Pittsburgh Steelers Football... Six Super Bowls and more.
Great Players in Pittsburgh Steelers Football ,,,Chuck Noll, Bill Cowher, Mike Tomin, etc.
Great Coaches in New England Patriots Football,,, Bill Belichick the one and only plus others
Great Players in New England Patriots Football… Tom Brady, Drew Bledsoe et al.
Great Coaches in Philadelphia Eagles Football..Andy Reid, Doug Pederson & Lots more
Great Players in Philadelphia Eagles Football Great players such as Sonny Jurgenson
Great Coaches in Syracuse Football All the greats including Ben Schwartzwalder
Great Players in Syracuse Football. Highlights best players such as Jim Brown & Donovan McNabb
Millennials are People Too !!! Give U.S. millennials help to live American Dream
The Candidate's Bible. Don't pray for your campaign without this bible
Rush Limbaugh's Platform for Americans… Rush will love it
Sean Hannity's Platform for Americans… Sean will love it
Donald Trump's New Platform for Americans. Make Trump unbeatable in 2020
Tariffs Are Good for America! One of the best tools a president can have
Great Coaches in Pittsburgh Steelers Football Sixteen of the best coaches ever to coach in pro football.
Great Moments in New England Patriots Football Great football moments from Boston to New England
Great Moments in Philadelphia Eagles Football. The best from the Eagles from the beginning of football.
Great Moments in Syracuse Football The great moments, coaches & players in Syracuse Football
Boost Social Security Now! Hey Buddy Can You Spare a Dime?
The Birth of American Football. From the first college game in 1869 to the last Super Bowl
Obamacare: A One-Line Repeal Congress must get this done.
A Wilkes-Barre Christmas Story A wonderful town makes Christmas all the better
A Boy, A Bike, A Train, and a Christmas Miracle A Christmas story that will melt your heart
Pay-to-Go America-First Immigration Fix
Legalizing Illegal Aliens Via Resident Visas Americans-first plan saves $Trillions. Learn how!
60 Million Illegal Aliens in America!!! A simple, America-first solution.
The Bill of Rights By Founder James Madison Refresh *your knowledge of the specific rights for all*
Great Players in Army Football Great Army Football played by great players..
Great Coaches in Army Football Army's coaches are all great.
Great Moments in Army Football Army Football at its best.
Great Moments in Florida Gators Football Gators Football from the start. This is the book.
Great Moments in Clemson Football CU Football at its best. This is the book.
Great Moments in Florida Gators Football Gators Football from the start. This is the book.
The Constitution Companion. A Guide to Reading and Comprehending the Constitution
The Constitution by Hamilton, Jefferson, & Madison – Big type and in English
PATERNO: The Dark Days After Win # 409. Sky began to fall within days of win # 409.
JoePa 409 Victories: Say No More! Winningest Division I-A football coach ever
American College Football: The Beginning From before day one football was played.
Great Coaches in Alabama Football Challenging the coaches of every other program!
Great Coaches in Penn State Football the Best Coaches in PSU's football program
Great Players in Penn State Football The best players in PSU's football program
Great Players in Notre Dame Football The best players in N.D.'s football program
Great Coaches in Notre Dame Football The best coaches in any football program
Great Players in Alabama Football from Quarterbacks to offensive Linemen Greats!
Great Moments in Alabama Football A.U. Football from the start. This is the book.
Great Moments in Penn State Football PSU Football, start--games, coaches, players,
Great Moments in Notre Dame Football ND Football, start, games, coaches, players

Cross Country with the Parents A great trip from East Coast to West with the kids
Seniors, Social Security & the Minimum Wage. Things seniors need to know.
How to Write Your First Book and Publish It with CreateSpace. You too can be an author.
The U.S. Immigration Fix--It's all in here. Finally, an answer.
I had a Dream IBM Could be #1 Again The title is self-explanatory
WineDiets.Com Presents The Wine Diet Learn how to lose weight while having fun.
Wilkes-Barre, PA; Return to Glory Wilkes-Barre City's return to glory
Geoffrey Parsons' Epoch... The Land of Fair Play Better than the original.
The Bill of Rights 4 Dummmies! This is the best book to learn about your rights.
Sol Bloom's Epoch …Story of the Constitution The best book to learn the Constitution
America 4 Dummmies! All Americans should read to learn about this great country.
The Electoral College 4 Dummmies! How does it really work?
The All-Everything Machine Story about IBM's finest computer server.
ThankYou IBM! This book explains how IBM was beaten in the computer marketplace by neophytes